Business Network Learning

Business Network Learning

Edited by

HÅKAN HÅKANSSON AND JAN JOHANSON

2001
PERGAMON
An Imprint of Elsevier Science
Amsterdam – London – New York – Oxford – Paris – Shannon – Tokyo

ELSEVIER SCIENCE Ltd
The Boulevard, Langford Lane
Kidlington, Oxford OX5 1GB, UK

First edition 2001

Library of Congress Cataloging-in-Publication Data
Business network learning / edited by Håkan Håkansson and Jan Johanson. – 1st ed.
    p. cm. – (International business and management series)
    Includes bibliographical references and indexes.
    ISBN 0-08-043779-6 (hardcover)
    1. Business networks.   2. Organizational learning.   I. Håkansson, Håkan, 1947–   II.
Johanson, Jan, 1934– III. Series.

HD69.S8 B865 2001
658'.044–dc21   00-051392

British Library Cataloguing in Publication Data
Business network learning. – (International business and management)
    1. Business networks   2. Organizational learning
    I. Håkansson, Håkan, 1947–   II. Johanson, Jan
    658'.044

ISBN: 0-08-043779-6

Contents

Contents

Series Editor's Preface

Inter-firm relationships have proved to be one of the most dynamic areas of business studies. A group of researchers from Europe – under the banner of IMP group – have been studying this field for the last three decades. These studies have resulted in several path breaking publications on different aspects of inter-firm relationships and have contributed towards the further development of theories in business research. The network approach introduced by this group in business studies is now an established field of research. The present volume looks at learning aspects of these business networks.

The present volume is edited by the two most prominent scholars from Uppsala school and looks at the learning effects of these business networks. Knowledge and learning acquired by firms through their internationalization activities and through relationships with other firms is in focus. It covers formal as well as less formal co-operative arrangements between companies, as informal business relationships are considered to contribute as much to the knowledge development of a firm as formal co-operative agreements.

As such this volume serves well to the main purpose of this series, which is to stimulate and encourage research on the new developments in the field of international business with special emphasis on marketing and management issues. We are thus very pleased to present this volume to our readers, as the most important contribution of this year and the very first volume of this century.

I would like to take this opportunity to recognize the professional assistance provided to us by Elsevier Science Limited (Pergamon) through their highly supportive team: Sammye Haigh, David Lamkin and Neil Boon.

PERVEZ N. GHAURI
Series editor

List of Tables

List of Figures

List of Figures

The Contributors

Maria Andersson   Department of Business Studies
Uppsala University

Ulf Andersson   Department of Business Studies
Uppsala University

Anders Blomstermo   Department of Business Studies
Uppsala University

Soon-Gwon Choi   Department of Business Studies
Uppsala University

Jonas Dahlqvist   Department of Business Studies
Uppsala University

Carin B. Eriksson   Department of Business Studies
Uppsala University

Kent Eriksson   Department of Business Studies
Uppsala University

Lars Frimanson   Department of Business Studies
Uppsala University

Amjad Hadjikhani   Department of Business Studies
Uppsala University

Håkan Håkansson   Department of Business Studies
Uppsala University and Norwegian
School of Management, Norway

Jukka Hohenthal   Department of Business Studies
Uppsala University

Ulf Holm   Department of Business Studies
Uppsala University and

The Contributors

	Institute of International Business Stockholm School of Economics
Christine Holmström	Department of Business Studies Uppsala University
Marleen Huysman	Department of Work and Organizational Psychology University of Delft The Netherlands
Jan Johanson	Department of Business Studies Uppsala University
Martin Johanson	Department of Business Studies Uppsala University
Katarina Lagerström	Department of Business Studies Uppsala University
Johnny Lind	Department of Business Studies Uppsala University and University of Manchester School of Accounting and Finance
Jan Lindvall	Department of Business Studies Uppsala University
Cecilia Pahlberg	Department of Business Studies Uppsala University
Ariane von Raesfeld Meijer	Department of Commercial Management and Marketing University of Twente The Netherlands
D. Deo Sharma	Copenhagen Business School Denmark
Lars Silver	Department of Business Studies Uppsala University
Torkel Wedin	Department of Business Studies Uppsala University

CHAPTER 1

Business Network Learning – Basic Considerations

HÅKAN HÅKANSSON and JAN JOHANSON

Introduction

The importance of learning

The business landscape is complex. Making up the landscape are business units intricately composed of resources with distinctive social, technical and economic properties. Every business unit combines certain hard physical resources such as facilities and products with soft resources, human or otherwise. These units are bound together in complex constellations or industrial structures where the resources are related to one another across firm boundaries. There seems to be general agreement that rapid internationalization and technical development are changing this already complex business landscape dramatically. Evidently, the complexity of the landscape itself, as well as the new developments, are closely related to knowledge and learning. There is also a striking double connection between the two, in that while new developments create a need for learning, learning is also a driving force in development. A similar relation exists between learning and complexity. That is, complex and intricate webs of resources require learning, and learning processes may help to create complexity.

When internationalizing, firms are exposed to new market situations and must learn about those situations. Both firms that internationalize operations and firms confronted by new companies and conditions in the domestic markets must learn how to handle new situations and developments that arise. At the same time, important learning processes may lie

1

behind the internationalization process. This can involve individuals learning new languages and social cultures, or organizations learning to operate in foreign countries. It can also mean exporting companies that learn about their foreign customers, or the reverse, companies learning about suppliers from abroad.

Similarly, new or changing technologies are in themselves the result of new knowledge and learning. Such changes also place firms in new market situations. Business market situations are continually being transformed by changing technologies, forcing firms to learn and adapt. Such changes may occur in production processes and products, and in relation both to suppliers and customers. New technologies must be learnt, though often the main impact comes in how they transform related technologies.

It is easy to understand the increased focus on knowledge development and corporate learning of the past decades as an effect of the increased interest in international and technological dimensions in business. Knowledge managers and life-long learning have become valid concepts; learning organizations, literature and research reports about learning topics abound.

The importance of business relationships

Another development, parallel to the increased emphasis on learning, has taken place concerning firm cooperations. In the middle of the 1980s cooperative arrangements between firms were introduced as major strategic issues and researchers gathered to analyze cooperative strategies (Contractor and Lorange 1988). Joint ventures, partnerships, licensing agreements, franchising, management contracts, strategic alliances and strategic networks came into focus. They have also remained there and the interest in them seems to be growing. A recurrent argument for this attention on cooperative strategies is that such strategies enable firms to become actively involved in the quickly accelerating knowledge development associated with internationalization and technical development (Beamish and Killings 1997).

While these formal cooperative arrangements are important and noteworthy, there is reason also to consider the less formalized, but close and long-lasting, exchange relationships between supplier firms and customer firms – *business relationships* – in business markets. Such business relationships are less recent than the cooperative arrangements mentioned above, and we find examples of them far earlier. In fact, they seem to be fundamentally associated with market economies. Less spectacular than the formal cooperative arrangements which have appeared on the scene more recently, these older, but also cooperative, business relationships are less illustrious in that they emerge out of ordinary market relations. Nevertheless, they have strong implications for firm learning and, depending on the extent to which this learning potential is exploited, can contribute just as much to the knowledge development of a firm as more formal cooperations.

In the European International Marketing and Purchasing (IMP) project, extensive empirical data on business relationships have been collected and analyzed (Ford 1990; Håkansson 1982; Turnbull and Valla 1986). The data demonstrate that most business firms are engaged in business relationships with a limited set of important customer firms. This set of customer firms accounts for a significant share of the business conducted by supplier firms. In addition, customer firms are considered important for the technical development of the supplier. Thus, the study shows that business relationships constitute a firm's business base. In this respect, there are no differences between the firms of the five countries investigated – France, Germany, Italy, Sweden and the United Kingdom.

The analysis also demonstrates that the relationships of the firms are, indeed, long lasting. The average age of business relationships at the time of the investigation was fifteen years. Transaction cost is one of the reasons for continuing business with a particular supplier. Production cost is another. It was stressed that close coordination of production activities provides firms engaged in long-lasting relationships with the possibility of lowering joint production costs. The case studies also show that close, long-lasting relationships contribute to more effective product and process developments than is otherwise possible.

Another finding of the IMP project is that relationships are established and developed through interaction between the firms. The relationship development process means that an initially weak interdependence between firms, associated with ordinary market relations, can be transformed into a strong and mutual dependence, which, in turn, allows the relationship partners to coordinate their interdependent activities and thus realize the gains mentioned above. Through interaction, firms are able to demonstrate their willingness and ability to do what they claim they intend to. This takes time, however, and therefore relationship development takes time. It also requires managerial efforts, in that managers, of various managerial levels and with different areas of expertise, must meet and exchange information repeatedly. The study shows that a great number of managers, in particular technical managers of various kinds, are involved in relationship interaction. Since each business relationship takes time to build and engages a number of managers, a firm's business relationships represent a considerable investment and have strong strategic implications.

A recurrent observation in the IMP study was that relationship interaction frequently is influenced by other relationships in which the interacting partners are involved. Customer's customers, supplier's suppliers, competing and complementary suppliers, consultants and intermediary firms can all have an influence on the interaction in the customer-supplier relationship. This gave rise to the general conclusion that relationships are connected to one another in the sense that the interaction in one has an impact on the other. Thus, each relationship is embedded in a set of connected relationships forming a network structure. It seems that business markets are networks of interconnected business relationships. In the following, we label such interconnected business relationships *business networks*.

Correspondingly, each firm is engaged in a business network structure of this kind, which contains both assets and liabilities to the firm. A firm's business network assets are associated with the firm's ability to continue doing business with some degree of predictability and without the cumbersome marketing and purchasing costs associated with being forced to switch to new customers or suppliers frequently. Business network assets also enable a firm to organize production more efficiently than would be possible if the firm had to form new cooperations with other customers and suppliers again and again. Moreover, assets provide the firm with a base for long-term development of products, processes and business in general. The liabilities of the business network embeddedness are associated with a loss of control over one's own business, which means that a firm may be forced to act in less efficient ways because the relationship counterparts demand it. Business network assets and liabilities both involve direct interdependencies with relationship partners and indirect interdependencies with the relationship partner's partners.

Since each firm is engaged in business relationships with a set of customer firms and supplier firms, which, in turn, are engaged in a number of business relationships, each firm is engaged in ever-extending business network structures. Thus, business networks are unbounded structures. When a firm interacts with its business relationship partners it modifies the network structure, which is also modified whenever other firms interact. Thus, we can say that, although the business network is stable, its structure is continually changing due to the interaction between firms. As a general rule, it can be said that the impact of changes in the network structure become smaller the further away from a firm they occur. It is also possible, however, that even distant changes to the network structure can have a radical and disruptive effect on a firm.

Business network learning

Learning in business relationships

According to organizational learning theory the critical issues in firm learning are associated with tacit knowledge, which can only be developed through experience. Such experiences are made when organization members interact with the environment. Learning occurs when the routines of the organization are modified. It is assumed that every firm can be viewed as a system of interrelated activities and that the routines of the organization are about the ways in which activities are performed and linked to one another. Thus, organizational learning means that those activity patterns are modified as a consequence of a firm's interaction with its environment.

Consider first learning in business relationships, the basic element in business networks. In fact, business relationships arise through learning processes. Ordinary market relations, in which there is no interdependency between two firms, gradually transform into business relationships in which the activities of the two firms are mutually dependent. If the process is broken down into phases, a first phase can be described as when firms learn about one another's willingness and ability to continue doing business together. This learning enables the firms to transact at lower cost than if they had been dealing in the ordinary market. This is a clear instance of learning that takes place through interaction. The two parties learn about each other and modify their routines for transaction thereby increasing the interdependency of those transaction activities. Obviously, such learning is facilitated through repeated interaction between the two parties over time. Moreover, we can also expect this learning to be stimulated by variation in the relation environment. Such variation exposes the parties to various pressures, which, if too great, can lead to termination of the relationship. If, however, the relationship survives such times of pressure, it will instead be strengthened, since the process teaches the parties more about how the other deals with different contingencies.

A second phase of learning occurs when two firms modify the routines that govern their production activities. Once again, it can be expected that such adaptations are based on experience gained through interaction. The learning in this phase may lead to the development of highly interdependent routines or even joint routines covering the activity systems of the two firms, thereby creating additional relationship value. In this case, the learning results in the joint performance of the two parties being higher than the sum of the performance of the two parties had there not been this joint learning.

In a third phase of learning in a business relationship, we can conceive of learning that leads to more long-term coordination of the activity systems of two parties, for instance, through development of new products or production processes. Once again, learning is based on repeated interaction and is stimulated by variations in the relationship and its context. Thus, the more managers involved in a business relationship's interaction and the more varied the expert input, the stronger the learning effects on the activity systems of the two firms. This kind of relationship may develop into quasi-organizations in which the activity systems linking two firms are more closely coordinated with each other than they are with the firms' other activities.

Above, we have discussed the three phases of learning in business relationships as if they follow each other in numerical order. This is not necessarily the case, and many firms go directly to coordinating production or organize joint development activities from the start. However, there do appear to be advantages in starting with small, routine adaptations which are sequentially and interactively developed, strengthened and extended so that the activities of the two firms become closely integrated. This can also be viewed as a social exchange process. Needless to say, learning in relationships does not always lead to closer relationships. It is perfectly conceivable that one of the parties – or both – discover that the routines required for further development of the relationship cannot be accommodated within their existing activity structures, or that a counterpart is not willing or able to make the modifications required.

In the literature on learning, it has been demonstrated that the effectiveness of learning depends on the absorptive capacity of the learning organization (Cohen and Levinthal 1990). Absorptive capacity refers to a firm's ability to value, assimilate and utilize external knowledge. It has also been shown that present and prior knowledge in a specific field has a strong positive effect on a firm's absorptive capacity. Thus, technical development of a firm is likely to occur in areas in which the firm already has considerable knowledge. Relationship learning, as discussed above, is an example of the importance of absorptive capacity as well as the development of it. Effective learning of partner firms about one another's abilities and needs is in part due to their already having a common knowledge base, which develops as a result of their interaction and mutual learning.

Learning from relationships

Thus far, the discussion on learning has only considered the learning that takes place within the relation between two parties. In a business network perspective, we have reason also to examine the effects of interaction in one relationship on learning in other relationships, that is, on how routines have a bearing on those relationships. Such effects may concern generalization and coordination.

One of the basic assumptions of the business network perspective is that each business relationship is unique. A business relationship has its own history and involves a specific set of individuals with their unique experiences and competencies. Nevertheless, there is reason to consider the possibility that experience from one relationship may also be applicable in another. We label this application of experience *generalization* of relationship learning. Examples of generalization can include establishment of relationships, response to initiatives by potential counterparts, engagement in joint production rationalizations, and product development cooperation. It is also conceivable that learning routines developed in one relationship can be applied in other relationships. Basically, it can be assumed that generalizations may be applied to relationships that are considered similar to those from which the experience was originally gained.

Experience from interaction in a business relationship can also lead to learning that develops links with other relationships. In any one relationship, a number of activities are performed, which are coordinated between the two firms. Since each firm is usually engaged in a set of relationships, however, there are also advantages to be gained by coordinating activities across relationships. This is often the case in just-in-time relationships where the customer firm must also coordinate with the just-in-time suppliers. Such coordination may entail several supplier relationships, or several customer relationships, or both supplier and customer relationships along a value chain. The development of routines for coordination across relationships may be a key type of business network learning. Cross-relationship coordination may be important for both similar relationships and those that are complementary to the focal relationship.

We have thus discussed learning based on experiences of interaction in one particular relationship and distinguished three kinds of learning: relationship development, generalization and coordination. But learning may also be affected by interaction in several relationships. It is generally acknowledged that variation or diversity stimulates learning. Firms that are exposed to a variety of relationships experience a wider range of relationship events. They are therefore likely to encounter more different kinds of demands and to experience more cases of failure and are therefore also likely to be more strongly motivated to modify and adapt routines.

Several organization-learning researchers have made the distinction between higher- and lower-order learning. Examples include Argyris and Schön's (1978) distinction between single-loop and double-loop learning, and Fiol and Lyles' (1985) lower and higher level learning. These and other related conceptualizations distinguish between learning that can be seen as modification, differentiation and specification of theories-in-use and associated routines, and learning that results in reconsidering and restructuring of existing theories-in-use and routines. The former type of learning takes place when experiences, on the whole, support existing routines and only call for minor amendments. The latter type of learning, on the other hand, occurs when a firm's experience forces it to question the basic assumptions on which strategies and structures are built. Exposure to a wide set of business relationships, that differ with regard to history, technology, culture and strategies, can be expected to lead to higher-order learning more frequently than does the absence of such exposure. However, as a general rule, it can also be expected that, for economic reasons, firms avoid higher-order learning as much as possible.

Levitt and March (1988) stress that learning occurs through one's own direct experience or from the experience of others. The discussions above concern direct own experience only. In the learning literature, it is often presumed that, in strategic alliances, learning from others is an important and interesting issue. In a business network perspective, the focus is somewhat different. Business relationships, and consequently business networks, are based on complementarity. Each firm in a business network specializes in a specific set of activities performed according to particular routines and utilizing a particular knowledge base. Each firm transacts and develops relationships with other firms, which specialize in other areas and the strength of business network structures lies in how routines complement one another and in the absence of overlapping knowledge bases. In this world, learning from the experience of others is an exception related to the unbounded nature of business networks.

Every firm has an interest in knowing something about firms other than those with which it enjoys relationships. Since learning only takes place through interaction with others, there are basically only two ways to learn. One is to interact with firms other than one's relationship partners. This is sometimes referred to as weak tie interaction and seems to have a role in learning innovative behavior (Granovetter 1973), such as moves into new technologies or new markets. The other method of learning is indirect learning, i.e., through the experiences of one's relationship partners. Knowledge of – in the network sense – distant parts of the business network is mediated through relationships. This allows firms to learn about network sections that they are unable to learn about directly. This kind of learning is not only mediated by intermediary firms, it is pre-interpreted and evaluated, and possibly included in the routines of those intermediary firms. It does, however, provide the firm with knowledge that can be used to modify various routines so that forces that impinge on markets and technologies, and which the firm has no direct experience of, may be accommodated in the firm's routine structure.

Relationships and increased opportunities to learn

Up to this point, we have seen the business world very much as a given and concentrated the discussion on how companies are able to use business relationships to learn about it. However, the business world is not at all a given, since the business units that make up the business world also develop it. Both internationalization and technological development are, at least partly, driven by the business units and how they combine systematically on the basis of business relationship development. Thus, the existence of relationships can be seen as an active force in the business world dynamics. One important reason for this is that relationships are part of the knowledge-generating process. Thus, business relationships do not simply facilitate learning, they also increase the number of opportunities to learn through expanding the total knowledge base.

From an economic point of view, the value of a resource is dependent on how it is used. Alchian and Demsetz (1972) conclude that this use – the value creation – is different for heterogeneous resources as opposed to homogeneous ones. The value of heterogeneous resources is dependent on the other resources with which they are combined. This is not the case for homogeneous resources. As discussed by Alchian and Demsetz, the most typical heterogeneous resource is that of human resources. When humans are combined, there arise special team effects due to how the involved persons affect each other. Thus, there is reason to combine humans systematically, i.e., a need for management. However, the process can also be looked upon from a knowledge standpoint, *cf.* our earlier reference to the importance of tacit knowledge. Much of the knowledge related to the effects of combining resources is probably tacit, at least at an early stage. Developing business relationships builds on combining different resource elements with one another, thus producing a continuous flow of new knowledge – as long as the resources are heterogeneous (Håkansson 1993). A substantial part of internationalization and technical development involves combining resources in new ways. This may include combining existing products with new user systems, or existing user systems with existing new production systems. In short, it is a matter of new combinations of single resource elements and new combinations of constellations of resources. The basis for this is already existing interdependencies or those that might be developed between any two resources. Combining two resources gives us an interface between the two. As we discussed above, this interface may be developed, or 'adapted'. The two resources can in any case not be used in isolation, but may, in turn, be combined with a third and a fourth resource and so on. Again there may be other interfaces developed that have an effect on the first interface. This means there are a large number of direct and indirect interdependencies between any two resources. The number of possible combinations is in fact so large – and continually increasing through new combinations with earlier unrelated resources – that it can never be completely known. Developing relationships thus produces an increasing total knowledge space that holds many opportunities for increased learning potential.

The layout of the book

Following this introductory chapter, the book is divided into three parts. The parts differ with regard to their focus on the business network learning processes as discussed earlier in the chapter. Part I focuses on learning in business relationships. Part II expands this view by discussing how embeddedness in the wider connected business network affects learning in business relationships, and Part III concerns learning in company networks in general, without consideration of specific business relationships.

Part I: Business relationship learning

Four chapters are dedicated to the discussion of various aspects of learning processes within dyadic business relationships. In Chapter 2, Håkan Håkansson, Marleen Huysman and Ariane von Raesfeld widen the perspective on learning by introducing the concept of inter-organizational teaching. It is argued that, in most relationships, inter-organizational teaching is equally important as learning and that studies of learning with no consideration to teaching will, in fact, oversimplify the picture of the learning process. Just as learning can be intentional or non-intentional, teaching can also be intentional or non-intentional. This distinction provides a typology which Håkansson, Huysman and von Raesfeld use in their discussion of a number of cases of inter-organizational interaction.

In Chapter 3, Lars Frimanson and Johnny Lind analyze the effect of the balanced score card on business relationship learning. Based on a discussion of strategic and organizational learning, Frimanson and Lind present an in-depth case study of the relationship between ABB and ASEA Skandia, concluding that the balanced score card maintains an internal focus which does not support inter-organizational learning as suggested by studies of business relationships in industrial marketing.

In Chapter 4, Ulf Andersson and Jonas Dahlqvist discuss how interaction in business relationships links sticky how-to-produce knowledge of the supplier firm, with sticky how-to-use knowledge of the customer firm, thereby promoting product development. In doing so, they suggest that knowledge evolves when actors from the two firms communicate openly and can demonstrate their understanding of the task at hand, as well as appreciate the conceptions held by other actors in the relationship.

A discussion of the importance of transferring knowledge to local units when firms go abroad follows in Chapter 5, where Soon-Gwon Choi and Kent Eriksson argue that, due to the specific nature of local markets, knowledge must undergo a translation process when being transferred from one national context to another. Choi and Eriksson analyze how this translation is brought about and how it is affected by various conditions. Particular stress is given the path-dependent nature of the translation process, and a case study is used to illustrate the problems associated with international knowledge translation.

Part II: Network relationship learning

The second part of the book recognizes that learning in business relationships does not take place in isolation from other relationships but is related to developments in the surrounding business network. The five chapters of Part II deal explicitly with these business network connections.

One problem of business network learning concerns the possibilities available to transfer knowledge from one relationship to another. It is generally stressed that each relationship is unique. This is the backdrop used by Kent Eriksson and Jukka Hohenthal, in Chapter 6, where they discuss the idiosyncrasies of business network relationships and develop a framework for analyzing the limits to transferability of relationship knowledge. In this discussion they place particular attention on the role of tacit knowledge in relationships.

In Chapter 7, Lars Silver and Torkel Wedin attend to the impact of different kinds of resource ties between companies in the development of new products. The case of a fuel engine development project carried out by Scania, a Swedish heavy truck manufacturer, and US diesel engine manufacturer Cummins is presented. Silver and Wedin also discuss how the interaction of this relationship is affected by Scania's relationship with Bosch, the German supplier of fuel injection systems, and of Cummins' other network relationships.

Chapter 8 explores the usefulness of business network experiences in the internationalization of firms. Kent Eriksson, Jan Johanson, Anders Blomstermo and Deo Sharma examine a firm's experiences with domestic and foreign suppliers and customers, as well as with customer customers and complementary suppliers, and how these factors relate to developing relationships with foreign customers. As could be expected they found that earlier foreign customer experiences are considered very important. They also find experience from other foreign relationship to be valuable due to its significant impact on the usefulness of experience gained from foreign customers.

In Chapter 9, Amjad Hadjikhani and Martin Johanson introduce expectations as a link between experiential knowledge and commitment decisions, with a discussion on how relationship and network expectations add a future dimension to the internationalization process model. Two longitudinal case studies of firm internationalization are presented to illustrate the role of expectations. Hadjikhani and Johanson find that general expectations, relationship expectations and network expectations all play a role in driving the internationalization process.

Earlier in the present chapter, we distinguished between three kinds of business relationship learning: relationship development, relationship generalization and cross-relationship coordination. In Chapter 10, Cecilia Pahlberg presents and analyzes a case of knowledge creation and diffusion in a multinational company (MNC) by starting from a strategic customer relationship of one of the MNC subsidiaries. Pahlberg demonstrates how this relationship is developed by learning, how the experiences from this relationship are generalized to other relationships, and how activities in other relationships are coordinated with this relationship. She also shows how knowledge gained from experience is diffused in the MNC network.

Part III: Company network learning

The third and final part of the book consists of three chapters on company network learning. First, in Chapter 11, Maria Andersson, Ulf Holm and Christine Holmström argue that competence development of MNC subsidiaries comes through relationship interaction. Based on this assumption, Andersson, Holm and Holmström analyze the roles of different types of relationships and their configurations, finding that both subsidiary market relationships and corporate relationships influence competence development. The relative roles of these relationships seem also to affect different subsidiary activities in different ways.

In Chapter 12, Katarina Lagerström discusses the use of cross-border and cross-functional projects as tools in the development of MNC networks. She examines how such projects can contribute to increased collaboration and, later, to the establishment of lateral relationships between MNC subsidiaries. Lagerström presents a case of an environmental management project at ABB.

In the final chapter, Chapter 13, Carin Eriksson and Jan Lindvall examine whether the greater need for coordination between subsidiaries within a network-oriented structure leads to the adoption of a management control system that creates better opportunities for organizational learning than the control systems in more hierarchical structures. An empirical study indicates that network organizations use non-financial and operational measurements and *ad hoc* reports slightly more frequently, though still to a very limited degree, than do traditional organizations.

References

ALCHIAN, A. A., and DEMSETZ, H., 1972, Production, Information Costs, and Economic Organization *The American Economic Review*, 62, pp. 777–795.

ARGYRIS, C., and SCHÖN, D., 1978, *Organizational Learning: A Theory of Action-Perspective*, Reading, MA: Addison-Wesley.

BEAMISH, P. W., and KILLING, J. P., (eds), 1997, *Cooperative Strategies*, San Francisco: The New Lexington Press.

COHEN, W., and LEVINTHAL, D., 1990, Absorptive Capacity: A New Perspective on Learning and Innovation *Administrative Science Quarterly*, 35, pp. 128–152.

CONTRACTOR, F. J., and LORANGE, P., (eds), 1988, *Cooperative Strategies in International Business*, Lexington, MA: Lexington Books.

FIOL, C. M., and LYLES, M. A., 1985, Organizational Learning *Academy of Management Review*, 4, pp. 803–813.

FORD, D., (ed.), 1990, *Understanding Business Markets: Interaction, Relationships, Networks*, London: Academic Press.

GRANOVETTER, M., 1973, The Strength of Weak Ties *American Journal of Sociology*, 78, pp. 1360–1380.

HÅKANSSON, H., (ed.), 1982, *International Marketing and Purchasing of Industrial Goods. An Interaction Approach*, Chichester: Wiley.

HÅKANSSON, H., 1993, Networks as a Mechanism to Develop Resources in Beije, P., Groenewegen, J., and Nuys, O. (eds), *Networking in Dutch Industries*, Apeldoorn: Garant.

LEVITT, B., and MARCH, J. G., 1988, Organizational Learning *Annual Review of Sociology*, 14, pp. 319–340.

TURNBULL, P. W., and VALLA, J. P., (eds), 1986, *Strategies for International, Industrial Marketing*, London: Croom Helm

Part I Business Relationship Learning

CHAPTER 2

Inter-Organizational Interaction and Organizational Teaching

HÅKAN HÅKANSSON, MARLEEN HUYSMAN and ARIANE von RAESFELD MEIJER

Introduction

During inter-organizational interaction organizations learn from other organizations (e.g. Powell 1998). Learning from others is a dynamic process that involves adaptation to the knowledge held by other organizations (Levitt and March 1988). Organizational learning is however just one side of the interaction; the other side is organizational teaching. Through the interaction an organization can influence others in a more or less systematic way. Although organizational teaching has not been given as much explicit research attention as organizational learning, we believe that the dynamics of organizational interaction cannot be studied and described fully when aspects of organizational teaching are left out of the analysis. In fact, we argue that organizational learning is hard to understand without bringing in teaching issues.

There are three main arguments behind our claim. A first one is simply that there is a lot of teaching going on. Our world consists of a number of large and/or powerful organizations that all try to control and direct the development of other organizations. Organizations invest billions and billions in marketing, in R&D, in technical development and in other means which are all directed at influencing others. Smaller organizations try to do the same, even if it is on a smaller scale.

A second reason is that without giving due attention to aspects of teaching, learning from others is too much seen as a simple autonomous process in which a single organization learns from the environment. Learning between organizations is seldom however a one-way directed process. The ecological character of learning has been pointed out earlier (e.g. Levinthal and March 1993, Levitt and March 1988). If we look not only at the learners but also at the role of the teachers during organizational interaction, we will gain a broader insight into the dynamics of the process.

A third reason to look at teaching while studying organizational interaction is to avoid a so called "active agency bias" that can be said to be present within our thinking of organizational learning. This bias refers to the tendency within the literature to see learning as an activity in which actors are more or less free to choose how to learn, what to learn, and from whom to learn (Huysman 1998). As such it refers to the assumption that learning agents are voluntaristic agents thereby overlooking issues of power that might influence learning. It can be stated that many authors who have analyzed organizational learning ignore issues of deterministic forces and consequently provide us with a rather romantic picture of an organization consisting of people able to "create the future" (e.g. Senge 1990). This romantic picture might change drastically when we include the role of the teacher in the discussion. Perceiving organizational interaction from both a learning and a teaching perspective reveals the issue of power that might influence the whole process of inter-organizational cooperation. Organizations are not at all free in deciding what, from whom and when to learn. The learning of an organization is often, in the same way as the learning of an individual, framed by some teachers.

In this chapter, we will make a first attempt to discuss the role of teaching during the process of organizational interaction. We will assume that teaching is a sub-process within the total process of organizational interaction. The ambition is to highlight the importance of teaching and try to identify when and how it influences organizational interaction and how teaching is related to learning. We will introduce a scheme for analyzing teaching situations. This scheme is based on the assumption that one major ingredient in both teaching and learning is the existence or non-existence of intentions. We will identify different situations of organizational interaction, related to the degree to which the two parties are aware of their role as a teacher or a learner, i.e. if they have the intention to play it or not. The various situations of organizational interactions will be described theoretically and illustrated by various empirical studies on organizational interactions. The paper ends with a discussion of the possible implications of looking at organizational interaction from a teaching point of view.

Organizational interaction, learning and teaching

The interaction going on between organizations has gained an increased interest during the last decades in organization research but even more so in economic studies on industrial markets. A large number of empirical studies have documented that the interaction between companies often develops into business relationships (e.g. Frazier *et al.* 1988; Gundlach *et al.* 1995; Morgan *et al.* 1994). Single business transactions are related to each other over time and result in relationships where the parties have feelings of both responsibilities, rights and obligations towards each other (see e.g. Ford (ed.) 1998). The content of these relationships varies and can be analyzed and assessed in different ways. Each relationship can include social, technical and economical elements. One suggestion is to analyze the content in terms of how closely the relationship links the two parties' activities to each other, to what extent it ties the two parties' resources to each other and to what degree it bonds the two parties as actors together. Furthermore, relationships are generally connected to at least some other relationships, which means that third parties are influenced. Thus, we can identify different forms of network effects on the development within one relationship (Håkansson & Snehota 1995). The most significant conse-quence is that organizations, through their relationships, become embedded into each other. Each organization is bound together with a set of other organizations. In the interface between the organizations, learning and teaching takes place. One important aspect of these organiza-tional relationships is inter-organizational learning. Given the ambiguity that surrounds the concept of organizational learning, it is necessary to state more explicitly to what process we refer when dealing with the concept.

We treat organizational learning as the process of organizational knowl-edge (re)construction. Emphasizing the construction of collective knowledge is in line with other recent contributions to the field that adhere to a constructivist perspective on learning (e.g. Brown and Duguid 1991; Cook and Yanow 1993; Nicolini and Meznar 1995; Pentland, B.T. 1995; Raelin 1997) and is inspired by the social constructivist approach to knowledge (Berger and Luckman 1966; Gergen 1994, Schultz 1971). Central is the way through which individual or local knowledge is "incorpo-rated" into collective knowledge or organizational knowledge. We refer to organizational knowledge as practices, procedures, stories, technologies, collective opinions, paradigms, frames of references etc., through which organizations are constructed and through which they operate. What is important is that organizational knowledge is independent from the single individuals. This is similar to the position of Attewel (1992) who argues that the "organization learns only insofar as individual skills and insights become embodied in organizational routines, practices, and beliefs that outlast the presence of the originating individual."

Basically, organizations learn in two ways: through their own experiences and through the experiences of other organizations (Levitt and March 1988). Learning through an organization's own experience includes experimenting as well as interpreting the results of past experience. Learning from others takes place through the transfer of knowledge of other organizations in the form of technologies, codes, procedures, or routines (Dutton and Starbuck 1978). During organizational interactions such as through inter-organizational networking, this adaptation to the experiences gained by other organizations becomes a key issue.

We perceive organizational teaching as the process of one organization sending signals with the actual result of (re)constructing knowledge of other organizations. By focusing on both the processes (sending signals) as well as the product of teaching (constructing knowledge of other organizations), we perceive teaching as a relational phenomenon. Consequently, because the existence and character of learning affects the teaching process in itself, we should always include learning when studying teaching.

Interaction situations

In order to make a first typology of interaction situations based on a teaching/learning perspective we will make use of the three key elements identified above: the teacher, the learner and the signal.

There are certainly a number of dimensions which can be used to categorize the teacher and the learner. They can be more or less competent, more or less well established, etc. In this first attempt we have chosen to concentrate on one dimension and that is the degree to which the teacher and the learner have more or less clear intentions during the process. The existence of clear intentions on both sides is certainly affecting the content as well as the effects. The importance of using intentions as a category has also been stressed by Finnemore (1996). She uses the concept of teaching to analyze the influences of international organizations on the creation of values. In the various cases presented in her study, she indicates that some organizations play the role of active teachers with well-defined lesson plans for their pupils.

We believe that there is a distinct difference between interaction situations where there are intentional teachers compared to situations where there are not. The same is believed to be the case also for the learner. Consequently, situations when both the teacher and the learner are intentionally active are different from situations where only one of the two sides has clear intentions. Furthermore, there is a clear difference between the situation in which an organization intentionally teaches an unintentional learner from a situation in which an organization learns in an intentional way from an organization that has no intentions at all to teach.

However, the existence of a signal in the process can start processes of teaching and learning even though there are no such intentions from any side at the start. The signal (e.g. a product) send from the teacher to the learner is a necessary ingredient in the interaction process. But such a signal also includes knowledge or value elements that might give an effect in terms of a (re)construction of the receiver's knowledge (i.e. learning), although it was not sent nor received with the intention to be enacted as such. Thus, a signal produced by someone without any teaching intentions is seen by someone else and is without any learning intentions reacted on and embedded into that actor. In such a case, learning has taken place and so has teaching.

In the next four sections we will discuss some typical situations based on the distinctions made above. Table 2.1 provides a scheme for analyzing the various interaction situations. Four different situations are identified. Each situation is illustrated with one or several empirical examples. Our intention is to demonstrate the differences in terms of the content of the processes but also of the effects of their generating differences in the teaching/learning dimension.

TABLE 2.1
A scheme for classifying interaction situations

	Intentional teaching	intentional teaching
Intentional learning	situation 1	situation 3
Unintentional learning	situation 2	situation 4

Situation 1: Interaction between two highly intentional actors

The first situation can be characterized as a typical educational situation. We have two actors with well-defined roles. One is seen both by itself and by the other as a teacher and the other in the same way as a learner. They agree that it is an educational situation and they are probably both aware of the need to create a good atmosphere for the interaction. The key question is to what extent their intentions overlap with regard to the subject.

Appointing a teacher

On a building site there are more or less continuously appearing needs of different products generally categorized as supplies, which can include tools, maintenance products, spare parts or minor products needed in the building activities. These needs can be handled in different ways. An expensive way is to let someone from the site go to a wholesaler or a distributor to fetch the needed products. An alternative is to try to work more closely with a distributor.

One construction company chooses this latter alternative. The two companies knew each other well already before this cooperation started and they were also situated close to each other. They initiated the process by making a first study in which they concluded that there were several problems. One was that the needs were difficult to forecast. Another that there was a large variation in what was needed and a third that these small purchases created a lot of administration. Together a solution consisting of three parts was designed. Firstly they jointly developed a catalogue in order to define the product mix. The catalogue consisted of three thousands items, being five percent of the total assortment. Secondly, it was decided that the distributor should mainly take care of the transportation. The latter had already a system of daily "trucking tours" and the construction company wanted to take more advantage of this. Thirdly, the companies agreed on a monthly billing system in order to reduce the paper flow.

However, one thing was to find the solution, another to get it to work. It took the companies more than five years to change the behavior of the persons involved. The construction company was highly decentralized and each of the site managers had to be convinced and so had all the different craftsmen who were responsible for different parts of the construction. The construction company wanted to use the distributor as a teacher telling the individual buyers and users how to behave. Whenever anyone wanted a solution that was outside what was agreed – if it was in regard to products or ways to handle them – it was the supplier's responsibility not just to inform but to try to persuade the individual buyer to stay within the agreed boundaries. In the end both parties perceived the project as successful (Gadde and Håkansson 1993).

There is a lot of "intentional inter-organizational education" going on. A large number of companies have their own "schools" where they train personnel working at their customers-organizations, to use, maintain, etc. the products they sell. Or the schools can be used to train suppliers to deliver the right quality at the right time. There are three possible outcomes of such intentional education. "Successful education" has occurred when the learner learns from the teacher as both parties intended. "Ambiguous education" has occurred when the outcome is different from what the teacher or the learner (or both) had in mind. "Unsuccessful education" has occurred when the education process does not result in any learning despite the intentions and efforts.

In cases where the teaching occurs intentionally and where the learner is aware of being taught, both parties might be better off when the organizational interaction is organized in a systematic educational way. By organizing teaching, actors try to limit the occurrence of ambiguous and/or unsuccessful education. This can take place for example through own schools as mentioned earlier, through special courses, through teaching contracts, or in the case of the construction company as described above, as part of an organized cooperation. This is a situation for managing teaching processes. There might be special managers (coaches) who have the right to intervene in the process so as to improve the outcomes. Many cooperation projects are, for example, managed in such a way as to support the process of knowledge exchange.

Situation 2: Interaction with a teacher being more active

There are two versions of a situation in which the teacher has clear intentions while the potential learner has not. One is when the teacher also has the ambition to make the learner become aware of its role – thus moving the situation to a situation with two intentional actors as discussed above. The second version is when the teacher has no intention of arousing such learning intentions – this situation could be described as more manipulative. Below we give two cases. One illustrates the situation of an intentional teacher who wants the learners to become intentional. The second case illustrates a situation in which the teacher does not want the learners to become intentional.

Teaching by "banging the drum"

Environtech, a Dutch organization for energy and environment, manages several programmes on sustainable development, energy-saving and environmental improvement. Through cooperation with industry, universities, government and the energy sector, Environtech tries to stimulate the development and application of energy-saving and environmental friendly techniques, technologies and instruments. In order to encourage energy and environmental innovation Environtech considers itself as an intermediary organization, which aligns research, institutes, producers, suppliers and users. The case describes in particular the activities of the programme warmth supply and how this programme manages its changing network in order to stimulate the use of warmth supply in housing. Residual warmth from the production of electricity or from the incineration of litter can be used for room heating. The use of residual warmth has a minor position in the Dutch energy supply. By providing knowledge, support and advice to energy firms, intermediary organizations and local governments on warmth supply, by developing a teaching programme on the subject and by participating in conferences and seminars on warmth supply, Environtech tries to stimulate the use of warmth supply. In principle the different activities of this programme represent different ways of teaching the actors in the 'energy' network.

Depending on a teacher

There are at the moment 572 social services (SSs) in the Netherlands whose main purpose is the provision of unemployment benefits to citizens

Håkan Håkansson, Marleen Huysman and Ariane von Raesfeld Meijer

within the municipality. As a result of a long established decentralization policy, the SSs are working independent from one another. The execution and administration of the provision of various social security services is highly complex and hence is traditionally being supported by computerized information systems. All SSs, apart from three larger cities, have outsourced their information system (IS) function as they have neither the budget nor the expertise to design and maintain these systems in house. There are at the moment five commercial software houses that are contracted to design and maintain IS for SSs. Each software house has its own group of clients varying from twenty-five to three hundred SSs. These relationships have a rather enduring character and sometimes exist for decades. The major reason for this continuance of the relationship becomes more understandable when we perceive the inter-organizational relationship between a software house and a social security as a teacher-learner relationship. Over the years, software houses gained in-depth knowledge concerning the administration and execution of service provision of a particular SS. Due to turn-over and lack of expertise, it happens frequently that software houses know more about the support of social services than a particular SS knows itself. Hence, software houses sometimes become the surrogate memory of the organization. Consequently, keeping in contact with the software house has become vital for many social services. The software houses are teaching the SSs about how to use the system but have no interests in teaching them about how they operate. It is better for them to keep the SSs depending on them (Huysman and Newman 1998).

The first case shows an example of intentional teaching in which the learners (the various partners in the energy and construction industry) are at first not aware of the existence and intentions of the teacher (Environtech) but in which the teacher wants to change the situation to one of intentional education. By banging the drum, the intention of Environtech was to create awareness among the potential learners of its knowledge and subsequently to "transform" them into intentional learners. The example shows that there exists something like a "teaching-role" – a position which at first might be empty but which can be filled by others. There is strong empirical evidence for the existence of teachers in terms of opinion leaders on the individual level (Rogers 1986). There are no reasons to believe that it is different on the organizational level – take for example the interest shown for lead users or for organizational knowledge brokers such as consultancy firms (e.g. Hargadon 1998) and the existence of 'promise champions' in the development of new technological domains, who need not be individuals (Van Lente and Rip 1998). Thus, there are good reasons for organizations to have the ambition to become seen as a teacher.

The second case illustrates a typical "marketing" situation. Someone wants to influence some passive "customers". The ambition is always to influence their behavior, sometimes through becoming intentional but often without that awareness. From the point of view of the teachers, it is sometimes more effective to have passive learners, as was the case between the software houses and the social security offices. In this case, the teachers have no intention to change the passivity of the learners so that they become active, intentional learners. From the (commercial) perspective of the software houses it is better to keep the situation as it is. Too much reflection on the situation from the side of the social securities would distract the current economic balance between the two parties. By facing passive learners, teachers can create a more stable, repetitive situation. In such a situation, there are no reasons to teach the learner so much that the teacher becomes redundant. In other words, it is more beneficial for the teacher to create single loop learning processes rather than double loop learning processes. During single loop learning the learner is given the solution without learning how to reach it, while during double loop learning governing variables that lead to the solutions are also being taught (Argyris and Schön 1978). However, interestingly enough, passive or single loop learning can also be beneficial to the customer, as Demsetz (1988) has argued. It is often more economical when the user can just use a product or a service without having to learn all about it. But it certainly gives the producer an advantage from a knowledge point of view.

Situation 3: Interaction between an intentional learner and an unintentional teacher

The third situation is the opposite of the second. The learner has the intention to learn but the teacher has no intention to teach. The learner is looking for a solution and tries to find this through a teacher. The teacher is unaware that it has a solution for someone else – that it has something to teach – or is uninterested in it. Also here we have two different versions. In the first one the learner has the intention to make the teacher intentional in order to change the situation into one of intentional education as discussed in section 4. In the second version there is no such intention. Again, two cases will be presented to illustrate the two situations.

Teaching a teacher to teach

A company within the pharmaceutical field developed a product that was used as equipment by hospitals. The system had been developed earlier but the company now wanted to design it in a more "industrialized" way. The ambition was to reduce the costs for production and at the same time get a flexible end product. It was thus a question of modularization. One important part of the system consisted of a pump. A buyer wanted the company to try to produce this in a new way. The supplier first perceived the idea as "totally unrealistic". The buyer's technicians had to be involved and after a period of intensive discussions the supplier became convinced that the suggestion was a new way of looking at the product. It started development work and managed later to produce the specified product (Gadde and Håkansson 1993).

In this case the pharmaceutical company had learnt about the product from the teaching supplier. The learner (having less experience within the field) could see possibilities by breaking with some of the established norms, and it could thereby formulate the product specifications in an innovative way. However, it did not have the knowledge or the competence to design or produce the product. It had to get the "teacher" mobilized and involved.

Learning through modeling

Over its years of existence, a specific work culture developed at the department of information system design (ISD) of AZ, a non profit company. To put it bluntly, IS designers perceived their job from a technological perspective and did not communicate much with other designers nor with the potential clients of the systems. Furthermore, the designers worked from nine till five and were not used to working overtime. In the beginnings of the nineties, the decision was made to commercialize AZ; it was time to revitalize the company, including the ISD department. Top management of AZ replaced the ISD department manager by a much younger and highly career minded manager who had a Ph.D. in information management. Unlike most designers who identified themselves with AZ, this manager identified himself more with the world of commercial software houses. By modeling the work of these software houses, he showed it was necessary to become more "cost-aware, client-friendly and commercially minded" and did not tolerate the nine till five culture. In his years as ISD manager, he tried to initiate organizational learning processes in which the IS designers learned to behave as designers working at commercial software suppliers. In other words, the IS manager acted as an intermediary in the teaching-learning process, while the actual teachers (commercial software suppliers) were not aware of being used as teachers (Huysman 1996).

This situation is interesting from a teaching point of view, as it probably is a rather typical starting point from which a new teacher is created. Signals produced by an organization – it may be a product, it may be a technique or it may be a way to organize or manage – can always be used by someone as a learning device. From a teaching point this situation is unproblematic, as the sending organization does not see itself as a teacher. But at the same time there is an interesting potential. There are always potential positive effects of organizations adapting to other organizations. We started out by describing how companies become embedded into each other, and this is one of the mechanisms. If the signals a company is sending out are so valuable that they are used by someone to learn from, there might be a development possibility to take advantage of. This was clearly the case in the first example in which the learner was eventually teaching the teacher. It also showed the complexity in the interaction in terms of learning and teaching. The two parties often changed between the role of teacher and the role of learner. In practice, the two roles exist side by side.

Benchmarking is another example of intentional learning versus unintentional teaching. The second case is an example of this. There is in this case no intention by the learner to make the teacher an intentional one. In fact, that could even destroy the possibility of using it as a model. In other situations organizations want to make the teaching organization at least aware of the fact in order to "reward" it. It is in general socially rewarding to anyone to be appointed as a "teacher".

Situation 4: Indirect interaction based on a signal

In the fourth situation neither the teacher nor the learner has any intention to teach or to learn. But there exists a signal – something sent out from the teacher and received by the learner – that has learning effects. It can be a product, a message or a routine. These signals have similar characteristics to viruses and genes, i.e. they are carrying something with them. As soon as they are received and brought into an organization they will create some effects. These effects are often minor, lasting only a short time, but there are exceptions. Some might have a severe effect, as will be discussed later. Furthermore, if a number of signals are carrying the same "disease", there might also be a cumulative effect. Each signal in itself does little harm but combined with others can give a larger effect. The key aspect is that an educational process is constructed out of the signals. Through that process some intentions may appear which then bring the situation into line with one of the three other situations discussed above. In other situations the process remains unintentional, both from the teacher's as well as from the learner's point of view.

Adopting technologies

An example which shows that the effects of indirect interaction based on a signal can be quite large is when an organization adopts a technology. Together with the physical substance, the organization in such a case also buys norms and values that are implicit within the technologies and of which both the buyer and the seller can be unaware. This is for example the case with computerized information systems (IS) that companies buy from software suppliers. Cultural aspects can always be found in computerized IS since such IS are ultimately a representation of reality and therefore also of a culture (Tibosch and Heng 1994). IS provide means of representing reality through a set of concepts and symbols, and in so doing, can be considered as a medium for the construction of social reality (Orlikowski and Robey 1991). Based on Giddens' "structuration theory", Orlikowski and Robey argue that ISs make it possible to institutionalize interpretation frameworks. An illustration of this process of internalization through ISs is offered by Walsham (1991) while referring to the implicit function of accounting systems. Accounting systems are predominantly used to set targets, to monitor performances and to identify and correct failures. However, these accounting systems are only one way of looking at the world that institutionalizes organizational boundaries and emphasizes certain numerical data. As such they can be seen as "institutionalizing the dominance of financial information" (Walsham 1991, p. 92).

This type of situation illustrates that learning and teaching might occur also in situations when there are no intentions. It appears due to the circumstances, due to chance. It indicates that a more conscious teaching/ learning process can start out due to a large number of reasons. One important consequence is the importance of the effects of all day-to-day interactions. All these mundane interactions, which are done again and again, have probably important teaching and learning effects. Firstly, they can give the start for more intentional processes but secondly, and maybe more important, they can be seen as some kind of combined "teaching and learning machines" successively embedding the two actors into each other from a knowledge point of view. They will give rise to educational effects on both sides, creating a joint world in which uncertainty is decreased but which also reduces room for individual action.

Concluding remarks

In this paper we have argued for the need to include the role of teaching when analyzing organizational interaction. Our ideas emerged from combining studies on organizational learning, inter-organizational cooperation and business-to-business relationships with each other. The analysis done leads to the following tentative conclusions.

Firstly, everyone probably agrees that all organizations learn. We would like to add that all organizations also teach. Some organizations do it extremely well and in a very organized way, others more unorganized. Some are not at all aware of it and still others have intentions but not the competence. A large number can probably develop their competence and skill in teaching. When we started the discussion we concluded that in order to capture teaching we had to include learning as it is an integrated part of the teaching. Given the statement above, there are probably very good reasons for including teaching as soon as anyone is interested in learning. Teaching is, at least in a large number of situations, an important part of the same interaction process on which the learning is based. Or as Finnemore (1996 p. 141) puts it, "The fact that policy makers spend so much time on rhetoric and on selling policies to publics, allies, and enemies is something that we should take seriously."

Secondly, our discussions as well as the cases have demonstrated the importance of intentions but maybe even more how important the interaction is for influencing the intentions. These are certainly not given once and for all by some internal features or values but are highly possible to influence.

Thirdly, our analysis has indicated the importance of interactions in themselves. Teaching is as all teachers know not just done in the classroom where teaching occurs formally. Instead, most of the teaching is done in day-to-day interactions, as parts of regular interactions. Thus, it is important that these day-to-day interactions are not separated and handled by some special staff. They should be seen as a major vehicle to influence all those close to the organization. In line with the notion of 'situated learning' (Lave and Wenger 1991) referring to collective learning processes that occur as part of regular work activities, it might be interesting to pay more explicit attention to "situated teaching" situations that are part of daily inter-organizational interactions.

Fourthly, just as most organizations would like to become so-called "learning organizations", it is not too far fetched to imagine organizations that would like to become "teaching organizations"; organizations that occupy a dominant position in their organizational field (e.g. Sahlin-Andersson 1996). However, organizations should be aware of the problems of being seen as too much of a teacher. Organizations become sometimes a role model for all others in relation to, for example, management issues. SAS was such a case in Sweden some years ago. This might be harmful as there is an obvious risk that the company becomes trapped in its own "success". Organizations might start to believe that they are the teachers in their field. But no one is just a teacher; we are also always learners.

Finally, the discussion also gives rise to considering the economics of teaching and learning. Learning and teaching have been described in terms of slow and fast adaptations (March 1991). There are clear economic reasons for the importance of being a fast adapting organization in certain situations but there are also economic reasons for being a slow adapting organization in other situations. One important example is the need to defend one's own investments such as in technical facilities, in order to use them in an economic way. This includes actively teaching others how to

Håkan Håkansson, Marleen Huysman and Ariane von Raesfeld Meijer

take advantage of these existing resources by, for example, adapting them
to their facilities and products.

References

ARGYRIS, Ch. and D. SCHÖN, 1978, *Organizational Learning: a Theory of Action-Perspective*, Reading, MA: Addison-Wesley.

ATTEWEL, P., 1992, Technology Diffusion and Organizational Learning: The Case of Business Computing *Organization Science*, vol. 3/1.

BERGER, P. and T. LUCKMAN, 1966, *The Social Construction of Knowledge*, London: Penguin Books.

BROWN, J. S. and P. DUGUID, 1991, Organizational Learning and Communities-of-Practice: Towards a Unified View of Working, Learning and Innovation *Organization Science*, vol. 2/1, pp. 40–57.

COOK, S. D. N. and YANOW, D., 1993, Culture and Organizational Learning *Journal of Management Inquiry*, vol. 2, pp. 373–390.

DEMSETZ, H, 1988, The Theory of the Firm Revisited *Journal of Law, Economics and Organization*, Vol. 4, No. 1, pp. 141–61.

DUTTON, J. M. and W. H. STARBUCK, 1978, Diffusion of an Intellectual Technology in K. Krippendorff (ed.) *Communication and Control in Society*, New York: Godon and Breach, pp. 489–511.

FINNEMORE, M., 1996, *National Interests in International Society*, New York: Cornell University Press.

FORD, D. (ed.), 1998, *Managing Business Relationships*, London: John Wiley.

FRAZIER, G. L., R. E. SPEKMAN and C. R. O'NEAL, 1988, Just-in-time exchange relationships in industrial markets, *Journal of Marketing* 52 (October) pp. 52–67.

GADDE, L-E. and HÅKANSSON, H., 1993, *Professional Purchasing*. London: Routledge.

GERGEN, K., 1994, Organizational Theory in the Postmodern Era in M. Reed (ed.) *New Directions in Organization Theory and Analysis*, London: Sage.

GUNDLACH, G. T., R. S. ACHROL and J. T. MENTZER, 1995, The Structure of Commitment in Exchange *Journal of Marketing*, 59 (January) pp. 78–92.

HÅKANSSON, H., 1987, *Industrial Technological Development. A Network Approach*, London: Croom Helm.

HÅKANSSON, H. and SNEHOTA I. (eds), 1995, *Developing Relationships in Business Networks*, London: Routledge.

HARGADON, A. B., 1998, Firms as Knowledge Brokers: Lessons in Pursuing Continuous Innovation, *California Management Review*, Vol. 40/3, pp. 209–227.

HUYSMAN, M. H., 1996, Dynamics of Organizational Learning, Amsterdam: Thesis Publishers.

HUYSMAN, M. H., 1998, Balancing Biases, A Critical Review of the Literature on Organizational Learning in Easterby-Smith, M., Burgoyne, J., and Araujo, L., (eds) *Organizational learning and the learning organization*, Sage Publications: London.

HUYSMAN, M. H. and NEWMAN, M., 1998, Developing Information Systems in a Turbulent Environment: the Case of the Dutch Social Security System Proceedings of the European Conference on Information Systems, June, Aix-en-Provence, France.

LAVE, J. and E. WENGER, 1991, *Situated Learning: Legitimate Peripheral Participation*, Cambridge, UK: Cambridge University Press.

LEVINTHAL, D. and J. G. MARCH, 1993, The Myopia of Learning *Strategic Management Journal*, vol. 14, pp. 95–112.

30

LEVITT, B. and J. G. MARCH, 1988, Organizational Learning *Annual Review Sociology*, 14, pp. 319–340.

LOOTS, J. and M. H. HUYSMAN, 1997, Exploring the Roots of Innovations, a Social Constructivist Perspective Proceedings of the conference on "Organizing in a multi-voiced world", Leuven, Belgium.

MARCH, J.G., 1991, Exploration and Exploitation in Organizational Learning *Organization Science*, vol. 2/1, pp. 71–87.

MORGAN, R. M. and HUNT, S. D., 1994, The Commitment-trust Theory of Relationship Marketing *Journal of Marketing*, 58 (July) pp. 20–38.

NICOLINI, D. & M. B. MEZNAR, 1995, The Social Construction of Organizational Learning: Conceptual and Practical Issues in the Field, *Human Relations*, vol. 48/7, 727–746.

ORLIKOWSKI, W. J. and D. ROBEY, 1991, "Information Technology and the Structuring of Organizations", *Information Systems Research*, vol. 2/2, pp. 143–169.

PENTLAND, B. T., 1995, Information systems and organizational learning: The social epistemology of organizational knowledge systems, *Accounting, Management & Information Technology*, vol. 5, 1–21.

POWELL, W., 1998, Learning from Collaboration: Knowledge and Networks in the Biotechnology and Pharmaceutical Industries *California Management Review*, Spring, vol. 40, no. 3.

RAELIN, J. A., 1997, A Model of Work-Based Learning, *Organization Science*, vol. 6, pp. 563–578.

ROGERS, E. M., 1983, *Diffusion of Innovations*, third edition, New York: The Free Press.

SAHLIN-ANDERSSON, K., 1996, Imitating by Editing Success: The Construction of Organization Fields in Czarniawska, B., and Sevón, G., (eds) *Translating Organizational Change*, De Gruyter Studies in Organization 56, Walter de Gruyter: Berlin.

SCHULTZ, A., 1971, *Collected Papers*, vol. 1 and 2, The Hague: Nijhoff.

SENGE, P., 1990, *The Fifth Discipline, the Art & Practice of the Learning Organization*, Random House: London.

TIBOSCH, M. J. M. H. and M. S. H. HENG, 1994, Information Systems and Organizational Culture: a View Point, *Malaysian Journal of Management Science*, vol. 3/2, pp. 17–33.

VAN LENTE, H. and A. RIP, 1998, The Rise of Membrane Technology: From Rhetorics to Social Reality, *Social Studies of Science*, 28/2, 221–54.

WALSHAM, G., 1991, Organizational Metaphors and Information System Research, *European Journal of Information Systems*, vol. ½, pp. 83–94.

CHAPTER 3

The Balanced Scorecard and Learning in Business Relationships

LARS FRIMANSON* and JOHNNY LIND**

Introduction

Recently, the Balanced Scorecard (BSC) has been introduced as a novel method for measuring performance in business firms, and it has gained interest among scholars as well as practicians ever since Kaplan and Norton published their first article in 1992. Since then, a number of articles and books have been published, further developing or analysing the BSC as a concept (e.g., Kaplan and Norton, 1992; 1993; 1996a; 1996b; Maisel 1992; Hoffecker and Goldenberg 1994; Olve *et al.*, 1997). However, a common strain in this literature is a weak empirical grounding, often trying to gain support through describing the process of implementation in various organizational units (Mouritsen *et al.*, 1996). This is not altogether surprising given the juvenile nature of the subject.

---

* Uppsala University
**Uppsala University and University of Manchester, School of Accounting and Finance.
The authors wish to thank Trevor Hopper, Håkan Håkansson, Ingemund Hägg and Jan Johanson for providing helpful suggestions and comments. We are also indebted to the staff and managers of ABB for making this study possible, especially to Mr Ludvig Bernhardsson at ABB Control AB. Financial support has been provided by NUTEK and Svenska Handelsbanken. An earlier version of this paper was presented at the workshop on New Directions in Management Accounting: Innovations in Practice and Research held by the European Institute of Advanced Studies in Management in Brussels, December 1998.

Even though Kaplan and Norton recognise the importance of the customer perspective, and that suppliers are a significant part of the internal perspective, little or no research has been performed on the BSC with respect to the consequences it has on customers and suppliers in an industrial market setting. This paper is intended to discuss the BSC and its consequences between two industrial companies. In particular, it focuses on how the BSC affects learning in business relationships, and it draws on the literature in industrial marketing and purchasing which has studied business relationships for a long period of time. Thus we interpret a relationship between two companies as an ongoing exchange process rather than in terms of single episodes and transactions. Håkansson and Snehota (1995, p. 25) describe this view as " [...] a business relationship is mutually oriented interaction between two reciprocally committed parties". As such, the business relationship is characterised as being long-standing between the companies involved, and taking place through several individuals in each company. Relationships often arise specifically to handle existing interdependence between two parties, but over time the relationships themselves will create new interdependencies between those involved. In this framework learning manifests itself by change. That is, when one party adapts something to the other, learning has taken place. Thus, learning can take place in various ways. It could, for example, occur through the modification of products or production processes, or by adapting an administrative system.

The purpose of this paper is to develop a model of the BSC and learning in business relationships and to compare and contrast it against Kaplan and Norton's notion of the BSC as a tool supporting strategic and organizational learning. The paper is structured as follows. The next section discusses the BSC, strategic learning and organizational learning. It is followed by a section providing a somewhat different view of learning through the concept of interorganizational learning in business relationships. The fourth section describes the empirical background of the BSC which is to be used as an empirical illustration, in turn followed by a discussion of the methodology. The sixth section is an empirical description of the firm's BSC in relation to its business relationships. This is followed by an analysis of its impact on interorganizational learning. The paper concludes with a discussion of the results and their preliminary implications.

Balanced scorecard, strategic learning and organizational learning

The BSC enables one to consider the performance of several aspects of a business. Kaplan and Norton (1996b) state four different perspectives that the BSC focuses on when measuring performance: learning and growth, internal business processes, customers, and financial outcome. Within each perspective, a set of measures is identified and objectives for each measure are decided. Kaplan and Norton also argue that there needs to be a cause-and-effect relationship between each separate measure within every perspective, further enabling perspectives to be linked to one another.

Relying on the idea of "what you measure is what you get" (Kaplan and Norton, 1992, p. 71), one fundamental feature of the BSC is its function as a tool, supporting and creating an environment throughout the organization that is conducive to learning. In particular, these authors advocate two distinct modes of learning: strategic learning and organizational learning. The first is mainly concerned with top management, and the second with organizational units throughout the firm. Within the strategic learning framework, the executive committee formulate strategies and then use the BSC to implement and receive feedback upon the chosen strategies (Kaplan and Norton, 1996a). This allows them to evaluate the effects of previously implemented strategies, in turn deciding whether to alter their plan of action or not. Thus Kaplan and Norton recognise that a conventional view of strategy may not hold for organizations of today:

> The strategies for today's information-age organizations, however, cannot be this linear or stable. Senior managers need feedback about more complicated strategies and more turbulent competitive environments. The planned strategy, though initiated with the best intentions and with the best available information, may no longer be appropriate or valid for contemporary conditions (Kaplan and Norton, 1996a, p. 251).

Consequently, Kaplan and Norton advocate that:

> Organizations need the capacity for double-loop learning, the learning that occurs when managers question their assumptions and reflect on whether the theory under which they were operating is still consistent with current evidence, observations, and experience (*ibid.*).

Kaplan and Norton's ideas coincide with those of Simons' (1990; 1991; 1995), who argues that the control system is a tool used both for capturing critical aspects of the business and for assessing changes of the planned strategy. As a consequence, according to Kaplan and Norton, the BSC creates opportunities for double-loop learning by being able to deliver information that may or may not question the strategy in use. As such, strategic learning means that the organization adapts its activities to the changed environmental conditions by making decisions about alternative ways of deploying the resources within it. This is what Simons (1995) denotes the interactiveness of control systems, the central theme of which is to capture new opportunities stemming from an uncertain environment, and which is mediated by managers further down the hierarchy.

Organizational learning, on the other hand, is conceptually encapsulated by one of the four perspectives in Kaplan and Norton's BSC. Three particular aspects affect organizational learning and growth: a) employee capabilities, b) information systems' capabilities, and c) motivation, empowerment and alignment (Kaplan and Norton, 1996, pp. 126–146). These three aspects are mainly concerned with matters internal to the firm. The underlying logic for commencing measurement in this perspective is to balance the behaviour induced by financial measures. When managers are evaluated on financial performance only, they tend to neglect the importance of sustainable investments, which are required to continuously enhance the capabilities of the employees, the information systems, and the behavioural processes in organizations. Employee capabilities are strongly related to the measuring of employee satisfaction, retention and productivity. The logic is that satisfied employees will remain in the organization, which, in turn, will affect productivity. The aim with this part of the BSC is to report the status of employee capabilities, and to determine whether investments are required for the re-education of the workforce, for the upgrading of the information systems that assist employees in their work, for boosting motivation and empowerment, or in the process of aligning personal goals with the goals set by top management through the BSC. Accordingly, the objective with this measurement structure is to report the status of the internal learning processes required for growth and for long term competitiveness. Thus, in Kaplan and Norton's model, organizational learning is created through the accountability function of performance measurement. Strategic learning, on the other hand, is more closely associated with the decision-making function of performance measurement.

Customer and supplier partnerships were recognised by Kaplan and Norton in their first article published in 1992. They argued then that the BSC supported the control of these partnerships. In their later work published in 1996, Kaplan and Norton have developed the customer and supplier aspects further. For example, in the customer perspective, customer retention and customer satisfaction are two of the core measures, and customer relationship is one of the value propositions they have identified. Consequently, the BSC is expected to measure aspects about customers and suppliers, and to create knowledge about these dimensions. However, their point of departure stems from the focal company's side only. This is evident from the illustrations given in the Rockwater case (Kaplan and Norton, 1993) since their illustrations do not focus on how it affects learning in business relationships. Kaplan and Norton do not discuss how the firm interacts with their customers and suppliers in order to influence the environment they are a part of. Instead, they view the firm as part of a market with a rather uncertain environment. In this framework, then, strategic learning focuses on how firms can adapt their strategy to a changing environment, and organizational learning is very much a question of how firms can learn to use their internal resources in a more efficient way.

Learning in business relationships

In the field of industrial marketing, an extensive amount of empirical evidence suggests that interactions between firms are characterised by long-lasting business relationships with a limited number of counterparts continuously adapting to each other's technologies (Håkansson, 1989; Hallén *et al.*, 1991). Thus, by and large, the business volume performed exists within these long-lasting relationships, which in turn have proved to be effective co-ordinating mechanisms for knowledge intensive exchange situations. As a result of this, studies have shown that product development, as well as technological development, is dependent to a great extent upon organizations' networks of business relationships (Håkansson, 1990). Furthermore, an important part of the learning function is to be found in the interface between firms, in which long-lasting business relationships not only create mutual demands on developing each other's capabilities in a particular dyad, but are also embedded in a larger network context since activities in a single relationship are not performed in isolation (Håkansson and Johanson, 1993; Anderson *et al.*, 1994).

The dyadic nature of the exchange performed in business relationships implies that both seller and buyer are of profound importance when assessing the relationship. Furthermore, the exploitation of resources in a relationship is so complex and intertwined that it is virtually useless to concentrate on just one party if it is to be evaluated. The relationship also develops over time, and its past history forms the basis for the present and the future. In addition, the relationship is contingent on the extent to which there is mutual recognition and understanding of the fact that the success of each firm depends in part on that of the other, with each firm consequently taking actions in such a way as to provide a co-ordinated effort focused on jointly satisfying mutual requirements. As such, the relationship is the means by which islands of knowledge, accumulated apart from each other, are gathered together and form the basis for new ways of combining resources.

In this context, then, no business is an island, and the conditions for learning are somewhat different than those guiding the BSC. Typically, Håkansson and Snehota assert that:

> [...] the organization is often embedded in its environment and that its behaviour is thus greatly constrained if not predetermined [by other organizations], which means that it is not a free and independent unit (Håkansson and Snehota, 1989, p. 187).

From this line of argument, Håkansson and Snehota propose a different set of conditions which exerts its influence on learning. Firstly, changes in the environment of the organization stem in part from forces inside the organization itself. The environment of an organization is not faceless, atomistic or beyond influence or control. Thus, business opportunities exist in the environment, and they are there to be discovered and exploited. Secondly, the strategy – the pattern of critical activities – of a firm does not entirely result from deploying resources that are hierarchically controlled. Resources can be obtained across organizational boundaries through means of exchange, and, consequently, neither the effectiveness of an organization nor its exchange possibilities are exclusively dependent upon the organization's relative efficiency at combining internal resources. Thirdly, executives in organizations are not the only group of employees that interpret environmental conditions and formulate, implement, and adjust strategies. Rather, this is performed by individuals throughout the organization, although coded and stored collectively. However, it is still the executives that are accountable for the results achieved through exchanges in business relationships. Accordingly, the management of organizational activities is interpreted somewhat differently within this framework:

> [...] continuous interaction with other parties constituting the context with which the organization interacts, endows the organization with meaning and a role. When this proposition applies, any attempt to manage the behaviour of the organization will require a shift in focus away from the way the organization allocates and structures its internal resources and towards the way it relates its own activities and resources to those of the other parties that constitute its context (Håkansson and Snehota, 1989, p. 198).

Learning in this framework, then, occurs through the interaction processes of individuals, and it is not easy to detach organizational learning from strategic learning. They seem to be intertwined. The organizational environment is also to some extent certain and possible to control through the learning processes that take place within the patterns of interaction in business relationships. Consequently, the distinction between what is internal and external to the organization is somewhat blurred. Rather, the locus for learning is in the interface, an interface which changes over time and space.

At least two differences can be identified between Kaplan and Norton's BSC and a BSC to be used in a business relationship setting. Firstly, there is a difference in the focus on learning and where it takes place. The business relationship framework focuses on the learning taking place between firms, whilst Kaplan and Norton's focus is on the learning inside the firm. Secondly, because the individual customer (or supplier) is known, and the firm has an established relationship with it, it is to some extent possible for the firm to control interorganizational learning. With these two observations as a point of departure, an empirical case will be used to illustrate differences between the two learning settings outlined above.

Background and ABB's EVITA scorecard

Asea Brown Boveri (ABB) employs 213,000 people in 1,000 companies in 37 business areas. In 1997, the ABB Group's sales were US$ 34,265 million. These business areas have been organized into four business segments. One of the four business segments is Industrial and Building Systems. It provides energy-intensive industries, such as chemical processing, steel making or the production of pulp and paper, and electrical systems inside industrial and commercial buildings, with the means to make industrial processes faster, more reliable, more environmentally safe and more energy efficient. The business segment Industrial and Building Systems employs 97,000 people world-wide, with 12,000 of them working in Sweden. In 1997 it had an annual turnover of SEK 117,963 million world-wide and SEK 18,149 million in Sweden. One business area within this segment is Low Voltage Apparatus, which develops and manufactures low-voltage (below 1000 V) apparatus for industrial and building installations. Production is located in six European countries with one of the production units being ABB Control in Västerås, Sweden. ABB Control is one out of some 100 ABB companies with a total of about 25,000 employees operating under the Swedish jurisdiction of the Group. ABB Control develops, manufactures and markets low-voltage products, for which it has world-wide product responsibility. Its products can be categorised into four groups: Control Gear, Switches and Breaks, Arc Guard Systems, and Programmable Logic Controls. In 1997, ABB Control's sales were SEK 600 million and they had 400 employees.

In the spring of 1994, ABB Sweden's Vice President of Finance, Peter Fallenius, declared the parent firm's intention of broadening the existing performance measurement system (Wennberg, 1994). The aim was to change the focus from financial evaluation to an evaluation process based more on operational performance in a decentralised setting. Consequently, Lennart Lundahl of ABB Management and Process Consultants was selected to be the director of the EVITA project[2].

Initially, according to Lundahl and Ewing (1997), EVITA's first phase was largely directed towards the conceptual development and design of a

---

[2]EVITA (Ekonomi och Verksamhetsstyrning I T-50 Anda) is an acronym of the project's Swedish name. The English translation made by Lundahl and Ewing (1997, p. 24) is: "business control in the T-50 spirit."

control model. This proved to be a difficult task because one of the main objectives was to design a scorecard that was inspired by Kaplan and Norton's BSC, but at the same time to re-evaluate the role of the traditional view of organizational hierarchy. Their argument for trying to decouple EVITA from the hierarchy derived from an earlier project within ABB Sweden, the so-called T-50 project. This project, launched by the former CEO, aimed to reduce processing time in all departments by 50%. One of its cornerstones was to decentralize decision-making and profit responsibility in order to enhance customer focus through flexibility, further enhanced by a competence development programme for all employees. Consequently, one of EVITA's objectives was to strengthen the T-50 message and further integrate it with other control systems, and not to design another control model "[...] for top management 'looking down' on the organization, whereas Evita's tool allows units on all levels to examine their own activities" (*ibid.*, p. 24). This means that the structure of measurement is based on each unit's tasks and operating procedures. And most importantly, and in contrast to traditional control systems, in EVITA it is not the measures that are broken down in the hierarchy, but the visions, tasks and strategies. In fact, EVITA measures were never intended to be aggregated up the hierarchy, which in turn means that organizational performance on a higher level is not measured by the aggregation of the performance of lower level units. Rather, the objective with EVITA was to provide each unit with a tool encompassing numerical values which can be independently evaluated and analysed at each organizational level.

The next phase was to choose the perspectives on which each unit had to build their EVITA scorecard, and the goal with this process was to provide the units with a balanced fundament of the responsibilities within the organization. And, different from the original blueprint, EVITA included five and not four perspectives that were subject to performance measurement: *innovation and development* to ensure organizational learning and growth; *employee competence and motivation* to be able to meet customer demands; *processes and supplier performance* to ensure smooth production; *customer satisfaction* to ensure quality of delivered products and services; and *long-term financial performance* to ensure capital supplier satisfaction (see Figure 3.1).

These perspectives are those that were decided centrally by the EVITA project group. On the basis of these perspectives, the actual process of forming the measures in each unit was arrived at by interpreting the unit's mission statement for each perspective. A process in which critical factors to the mission are defined, as well as the actions that the unit is required to undertake in order to be linked to the critical factors. Finally, according to Lundahl and Ewing, this allows each unit to isolate specific measures that are in conjunction with the universal mission statement. In addition, it was also centrally decided by the EVITA project group that each and every

FIGURE 3.1

The logic underlying the measurement structure of EVITA. Adopted from Lundahl and Ewing (1997)

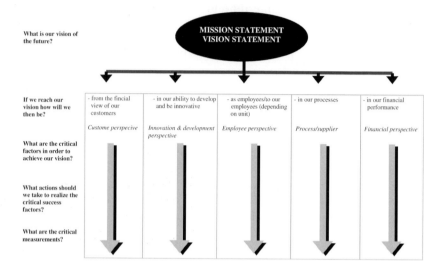

perspective was to have a maximum of five measures for simplicity. Two pilot companies started to implement the EVITA scorecard in 1995, ABB Coiltech and ABB Control. The department of logistics (hereafter Logistics) at ABB Control (hereafter ABB) is the unit of investigation in this study.

Method

A number of working papers, project reports and articles about EVITA have been produced ever since ABB launched the EVITA project (e.g., Ewing, 1995; 1996; Ewing and Lundahl, 1996; Bengtsson, 1997; Lundahl and Ewing, 1997). Fortunately, we have been able to discuss EVITA with ABB staff on several occasions. Lennart Lundahl and managers in various ABB companies have been kind enough to listen, discuss and explain their views about EVITA at workshops, firm presentations and seminars that we have participated in. Also, we have been able to recruit EVITA staff as well as ordinary EVITA users as visiting lecturers over the past two years.

The study is based on empirical data, applying a site visit method. In addition to the encounters above, data have been collected using semi-structured interviews over a two-day period. The information collected derives from five interviews ranging from one to three hours. The interviewees consisted of the managing director, the chief accountant and, at Logistics, the head of department, one purchaser and one person working with customer service. In addition, the study relies on internal documents as well as post-interview telephone conversations and e-mail communication for interpretative adjustments. Quotations noted in the study are drawn from the interviews accordingly, although some interviews were recorded and some were not.

Empirical description

Business relationships

Although driven both by inventories and customer orders, the manufacturing process at ABB is mainly comprised of small batch sizes. ABB has some 1200 customers, some which have remained stable during the last five years. Approximately ten customers represent 50% of the sales. ABB has explicitly tried to change its focus towards customers. Previously they had been more concerned about focusing on products rather than the processes surrounding them. Having determined that customer drop-out was often caused by errors in relating processes, such as logistic activities, rather than those attributable to the product, ABB has been trying to deliver total concepts encompassing order systems, customer service, transport systems and the like. The overall objective is that customers will perceive them as the most cost-efficient supplier. In doing so, ABB has prioritised three customer segments as being particularly important: original equipment manufacturers, relay interlocking plants, and wholesalers, together with their customers.

Since the technical aspects and the price are similar to those of the competitors, and after acknowledging that customer drop-out is caused by not providing appropriate services alongside the products, ABB seeks to establish partnerships with selected customers to learn about them and their related activities to increase future customer value. As such, developing co-operations with suppliers is perceived to be important to improve processes and cut costs, and to develop EDI/Internet order systems. Product development, although not in focus in this study, is aimed at quality assurance during early development phases (keeping in mind that the product concerns voltage apparatus), but also to develop 'smart' apparatus that can be contacted by making a simple phone call and asking about its status. A key role in their ability to increase customer value is identified with the daily conduct of employees. In terms of employeeship, ABB seeks to develop the competence of all employees, explicitly stating that all employees are a part of ABB's overall business development. This means creating a working environment comprised of safe and stimulating jobs, having individual 'competence plans', and developing visionary leadership among all managers. A key theme that has been stressed is to let employees come forward and to give them freedom to act.

ABB's largest customer in terms of sales is ASEA Skandia AB (hereafter Skandia), a wholesaler of installation materials used by electricians. ABB's relationship with Skandia can be characterised as an institutionalised business relationship. In fact, Skandia used to be a totally integrated part of the ABB Group, but today it is a completely independent firm. All in all, Skandia has an annual purchasing volume of MSEK 60 from ABB. Skandia has outlets all over the Nordic countries, but ABB mainly delivers to their distribution centre located in Örebro. The products delivered are components and larger units of assembled components, e.g., electrical fuse units, and most are standardised through rules governed by the CE-marking system, which is the quality system adopted within the European Union. As a wholesaler, Skandia's operations are driven by inventories. By and large Skandia's suppliers and their vast numbers of customers reside within Sweden. However, one of Skandia's largest customers, Ericsson, buys products to be incorporated in their cellular systems all over the world. In fact, ABB often participates when Skandia is discussing product and process development with Ericsson.

ABB's sales volume to Skandia has stayed more or less at the same level for the last five years and ABB supplies approximately 50% of Skandia's needs. Delivery takes place on a daily basis. Further, ABB's obligations towards Skandia have increased during the same period. This can be demonstrated through their business relationship which, today, includes several aspects related to mutual adaptation and the development of products as well as processes. One example is that they organize an annual marketing event together aimed at Skandia's customers. Other adaptations made by ABB include delivery of a special product without any packing whatsoever, and one concerning the maintenance of Skandia's item numbers. In addition, two major product development projects are operating today, although they are focused more on standardisation issues than on technical problems. In fact, the co-operation between the two firms has increased during recent years since both of them are aware of the benefits involved. Perhaps the most evident outcome of recent co-operation is a new order system based on Internet technology. It not only allows Skandia's customers to place orders through the Internet via a WWW site, but also enables them to obtain information about the delivery status. This means they can do this from home-office, construction site or wherever they can use a computer with a cellular phone. The order is then transformed via an EDI-box located at ABB, thereby establishing an interface to those working in manufacturing as well as customer service, and subsequently produced or taken from stock, depending on which ABB perceive to be most beneficial.

A direct result of the co-operation between the two is that ABB has been able to reduce production costs and increase sales. For its part, Skandia has also been able to reduce costs, but the main advantage that it has seen is an increase in knowledge. One example of this is that many Skandia employees have participated in the ABB Control School, aimed at increasing know-how about security issues related to handling and installing. Thus, Skandia employees are now able to discuss security matters with their customers, a vast number of whom are electricians.

The design and use of EVITA at Logistics

Logistics employs ten people: one head of department, three purchasers, four working with customer service (including the receipt of orders) and two working with back-office tasks. Logistics is one of six departments within ABB that fully operates with an EVITA scorecard.

The implementation process at Logistics started with a discussion about the department's vision for the future. Those involved in the discussion were the three purchasers, the four working with customer service and the head of department. The starting point when the vision was discussed at Logistics was one formed by the parent firm's vision, which had been communicated in a strategy document compiled by ABB and named Strategi 1998. Consequently, the vision on which Logistics built their EVITA scorecard focused on aggressive process improvement towards becoming "world-class", identifying customer value, order/logistics, product development, and employeeship as critical success factors.

On the basis of these critical success factors, EVITA project staff from the parent firm asked the employees and the head of department at Logistics: "If you are to excel in these areas, what will it take to do it?" Logistics discussed this internally and came up with their departmental success factors critical to the goal of becoming world-class, namely: the back-up function, accessibility, agreements with suppliers that are in conjunction with actual performance, information systems that support the logistics of both customers and suppliers, communication amongst employees and between the head of department and the employees, and finally, the work of economising on the inventory levels and capacity in relation to orders received. On the basis of these success factors, Logistics perceived it as being a rather simple matter to select the measures that would capture the critical success factors. After that, the head of department decided who would follow up each of the measures, taking into account those with particular problems in that respect. For example, one employee working with customer service had a number of customers in a certain region with voided contracts, and, consequently, she was selected to monitor, maintain and report the status of voided contracts for the entire department.

To summarise, the five perspectives and their measures of critical success at Logistics are as follows: *in their ability to develop and be innovative* they measure the number of active projects with suppliers and the number of sales calls at prioritised customers; *in the view of employees* they measure

Lars Frimanson and Johnny Lind

the number of weekly meetings conducted according to schedule, employee satisfaction and number of hours spent on employee education; *for their internal and supplier processes* they measure the share of orders placed by customers with valid contracts, lead-time from suppliers, degree of EDI order confirmation from suppliers, and level of service performed by suppliers; *from the view of their customers* they measure accessibility to technical back-up, level of service in relation to confirmed service time, level of service in relation to requested service time, and degree of EDI usage in customer relationships; *in the view of their financial performance* they measure inventory SEK per invoiced SEK and the number of orders received (see Table 3.1).

TABLE 3.1
EVITA perspectives and measures at Logistics

Innovation & Development	Employee	Internal & supplier process	Customer	Financial
number of active projects with suppliers	number of weekly meetings conducted according to schedule	share of orders placed by customers with valid contracts	accessibility to technical back-up	inventory SEK per invoiced SEK
number of sales calls at prioritized customers	employee satisfaction	degree of EDI order confirmation from suppliers	degree of EDI usage in customer relationships	number of orders received
	number of hours spent on employee education	lead-time from suppliers	level of service in relation to confirmed service time	
		level of service performed by suppliers	level of service in relation to requested service time	

Purchasers working at Logistics are heavily involved in calculations of the inventory levels and other logistical problems, but not so much with setting standard prices, which is a task mainly performed by the product managers. Accordingly, the decentralised purchasing function is a result of a policy decision taken within the business area, and is perceived as having good and bad aspects. It is beneficial in terms of flexibility and short communication channels, but at the price of an overall decrease in the purchasing base. The purchasers are also responsible for one or two inventories apiece, and much of their attention is aimed at ensuring an economical and steady flow of goods.

The purchaser interviewed stated that the most important aspects for tasks to be performed successfully are: 1) having the correct products in the inventories; 2) good relationships with suppliers; and 3) support from managers and colleagues. To ensure that the correct products are in the inventories, the purchasers use forecasts from a purchasing/logistics system, the PEO-system (Planning-Economy-Order), which provides inventory related information, such as the minimum, maximum and buffer levels, and also information about lead-times towards customers. This information is updated continuously by information supplied from the sales staff, and it helps the purchasers to make judgements about the inventory status and whether to purchase or not. An important task for the purchasers is to perform maintenance on the PEO-system, making sure it is run by the "right variables". In fact, purchasers often discuss this matter with sales people, and the purchaser interviewed in this study confirmed this by saying: "[…] shit in to the system means shit out of the system."

Good relationships with suppliers are controlled in two main ways: using the PPC-system (Partnership Performance Criteria) for evaluating supplier performance using some measurements, and through contact by telephone, at meetings and electronic conferences. The PPC-system is perceived as an important source of information, delivering information about the total cycle time and on time delivery, that is subsequently transmitted to EVITA. On time delivery is measured both as *orders delivered to ABB* and *orders delivered to customer* (for products that ABB buys and does not fabricate), and quality is measured in terms of scrap rate and through non-conformity reports. In fact the information delivered by the PPC-system was considered so important that a special OTODD (On Time One Day Delivery) programme was launched to help improve the PPC measures. The OTODD programme measures the performance of the PPC-system, ensuring it is used actively and improved. Furthermore, the OTODD programme is also used to register the number of meetings with suppliers.

Yet another important aspect when controlling purchasing tasks is the support gained from managers and colleagues. In particular, upon implementing EVITA, they began to measure the number of hours spent on product education with product managers. As asserted by the purchaser, this has really helped them: "We have actually improved a lot. Before, it could be like – Oh! Do we have another new product?" Today, before each product release, almost every employee at Logistics participates in the product education programme, which in turn enables them to update their respective systems accordingly.

Similarly, the employee working with customer service perceived that if one were to perform tasks successfully, the most important aspects were those stemming from customers. This involved ensuring that there was someone to answer questions, handle claims or deal with other irregularities smoothly, or just be in place to answer phone calls or e-mails. Consequently, feedback from customers is important, and it is nearly always instigated by the customer. In fact, the employee working with customer service asserted that she had several daily phone conversations with Skandia. Besides the phone, in order to control the relationship with Skandia, customer service employees use the PEO-system to obtain information about the level of service in relation to confirmed service time and number of orders received, which are the two measures perceived to be most important for controlling customer relationships. Interestingly enough, both these measures are used in EVITA, but there is one major difference with the PEO-system: it is updated continuously, whereas EVITA is only updated once a month. In fact, number of orders received is perceived to be so important that employees check it several times per day. And with the measuring of level of service in relation to confirmed service time, they can control what has been delivered or not for each customer, and they have the possibility of altering the order of delivery, in favour of prioritised customers, for example. This course of action is not possible with EVITA since it aggregates all customers to one level of abstraction.

Not to forget, explicitly confirmed by the managing director and the head of department, much of the control within the organization and outside it, in terms of relationships with suppliers and customers, is performed face-to-face. Intraorganizationally, this takes place using management by walking around and with scheduled meetings on a weekly basis, meetings which Logistics began to measure when EVITA was implemented. In fact, all interviewees at Logistics mentioned the positive results of beginning to measure the number of weekly meetings conducted according to schedule, resulting in all employees and the head of department knowing there is a weekly forum where all participants can address problems or aggravations. Interorganizationally, face-to-face control takes place through meetings with suppliers and customers, of which EVITA began to measure the number of meetings with prioritised customers. Further, the head of department also receives a separate measure of the frequency of complaints, which is perceived by him as being important: "Deliveries can never be too good! Only more or less acceptable." However, this measure has not yet been integrated in EVITA.

In terms of performance evaluation, both employees interviewed stated that they perceived being evaluated mainly on their abilities to communicate and do business with their counterparts. In fact, both the head of department and the managing director stated a clear customer focus, both when they were making the evaluation and when assessing what they thought was important when they themselves were subject to evaluation. However, this is not surprising with respect to the ABB Group's visions of achieving "world-class" performance, visions that were used when designing the EVITA scorecard. Thus, EVITA is used, although not directly, as a tool for evaluating their work towards becoming world-class. In the case of the managing director, EVITA mainly provides him with information about departments' work with employeeship, customer related information and internal processes. This information enables him to not only guide and control operations, but also to support department managers in their tasks. The managing director asserted that they are about to implement a corporate database, which will enable him and others throughout the company to monitor EVITA performance from each department. Using the same metaphor, both the head of department and the managing director stated that: "I mainly use EVITA as a clothes hanger on which to hang our operations. It is a tool for creating awareness."

To conclude, much of the vital information used when controlling Logistics business activities is derived from input systems prior to EVITA in terms of up-to-dateness and the frequency of reporting. Thus, EVITA is mainly the secondary source of information, although some new measures, e.g. number of weekly meetings conducted according to schedule, have been asserted as important. However, as the head of department stated:

> Although its high level of aggregation does not make it possible to evaluate our performance with Skandia, it certainly affects our consciousness about some aspects concerning customers, which in turn will indirectly affect our relationship with Skandia. After all, if we perform well for all customers, we will also perform well for Skandia.

EVITA and learning in business relationships

It is evident that managers and front-line employees at ABB do not use their scorecard when running daily, weekly or even monthly operations. None of the interviewees stated that they use EVITA directly for making decisions or that it enables them to identify and direct their attention to problem areas. The simple answer is that they use information from other sources which provides them with more accurate and relevant information. Some of these sources are the formal information systems which may be denoted as primary systems since much of the information produced by them is also used when loading information into EVITA. And it is the information delivered by the primary systems that employees primarily use when doing business with customers and suppliers, and which enables them to identify problems and make decisions.

The managing director and the head of department emphasize that they primarily use EVITA to obtain a more thorough picture of the development of activities at the departmental and firm level. This indicates that EVITA is mainly concerned with the function of managing and evaluating performance and, eventually, strategic decision-making. Thus, the focus is on accountability rather than operational decision-making. Also, the managing director's intention of implementing a centralised database encompassing all the departments' EVITA scorecards further strengthens the notion of accountability. That is, although it was not the intention when EVITA was created, it actually operates as some kind of hierarchical control mechanism; a control mechanism advocating organizational learning through the function of accountability within ABB.

Prior to the implementation of EVITA, the accountability function within ABB was mainly governed by the financial reporting system, ABACUS, which was built up from the smallest unit held accountable within a department all the way up through the divisions, to the corporate level and further up in the Group. At that time any upper organizational unit mainly paid attention to reports concerning financial items, and any unit doing the reporting devoted time and effort to delivering financial results that hopefully would be perceived as satisfactory.

From the discussion above, two scenarios can be depicted. The first concerns a reduced degree of freedom for the employees at Logistics because of the more comprehensive and penetrative accountability function. This is possible since corporate management may focus its attention to a greater extent on whether Logistics will attain the performance targets set through EVITA. Prior to EVITA, the accountability function was mainly concerned with whether Logistics had achieved its financial targets or not, and not with how to operate in order to achieve these targets. Consequently, the implementation of EVITA resulted in the corporate management being more concerned about evaluating a number of non-financial measures in four additional perspectives, limiting the means by which one could demonstrate that one had attained the financial targets. Thus, the narrowing of employees' degrees of freedom to act also limited the possibility of obtaining a climate conducive to learning in business relationships. This occurs when employees are not able to interact with a certain customer or supplier without being concerned about how their mutual actions for adaptation will affect the outcome in the non-financial structure of measurement. From a control point view, of course, the question is for whom, with what and to what extent the alignment of individual and organizational goals should be made. Support for this scenario can be found in the literature on learning through local information systems. Several empirical studies have shown that local information is an essential ingredient for local learning to take place (e.g. Jönsson and Grönlund, 1988; Grönlund, 1989; Solli, 1991; Westin, 1993; Jönsson, 1996). As proposed by Mouritsen *et al.* (1996), new accountings, such as the BSC, entail a transition from a flexible business setting to a stabilised business process re-engineered setting, which inhibits the possible ways of 'doing business'. The BSC's context is denoted by top level strategy and not on the simple day-to-day learning that takes place through local individuals between organizations. The quest lies in whether or not the BSC supports the standardising and centralising efforts made by the top level strategy, or if it supports the unique and historically determined conditions conducive to interorganizational learning and decision-making.

The second scenario depicts EVITA as a means for a more formal relationship of accountability between ABB and its customers and suppliers. If EVITA can be used as a tool conveying an active dialogue between two equal partners, who define the aspects they consider to be of importance together, the focus of accountability will change from an intraorganizational and hierarchical focus to an interorganizational focus distinguished by mutual adaptation and commitment. Thenceforth, commonly decided aspects in a business relationship are most likely to result in both parties being concerned about learning that is relevant to these particular aspects. And the degree of interorganizational learning in these predefined aspects will increase since the parties involved will put time and effort in the horizontal entity of accountability. However, one must remember that interorganizational learning takes place in various ways, utilising accountability in business relationships being only one of many. So defining these interorganizational relationships of accountability simultaneously means the exclusion (or, at least, the diminution of the importance) of other learning interfaces between organizations. Still, a hypothetical example of this can be drawn for this case from the co-operation in developing the Internet order system with Skandia. If both Skandia and ABB mutually decide to measure the degree of Internet orders placed by Skandia and its customers, a faster rate of learning is likely to occur.

Discussion

The conclusions drawn from the site visit method used in this study should be considered to be a first exploratory step in the analysis of the BSC and its consequences for learning in business relationships. Håkansson and Snehota (1995, p. 201) assert that: "Everything is possible if an actor gets the support of the network, while the same time nothing can be done if the network goes against the actor." The key then is to utilise learning (and perhaps teaching) processes through business relationships to obtain the support of the network. Thus, the business relationship approach is focused on how to externalise the internal capabilities, and is directed at analysing learning between companies.

In contrast, despite Kaplan and Norton's propositions about customer relationships, the main task for the BSC is to internalise the external environment. In this framework, then, learning has an internal focus. Thus the intended role of the BSC is, primarily, to support strategic decision-making, and to establish internal relationships of accountability that reflect the capabilities (i.e., resources) of the organization. As such, its role is to control the organization's efficiency in its process of combining internal capabilities, which in turn determine its effectiveness in the external arena denoted as the environment. Consequently, within this framework, strategic learning is closely related to adapting to the environment.

However, if one applies the framework provided by scholars concerned with industrial markets, learning is more a question of enacting among known individuals and organizations. In this framework the distinction between strategic and organizational learning is not particularly evident. Rather, some of the sources of effectiveness stem from the internal capabilities of the organization and the formulation, implementation and adjustment of strategies are also performed by individuals at all levels throughout the firm, and not only by executives. Consequently, the external is a part of the internal and *vice versa*. The question then is how an internalising control system such as the BSC affects an interorganizational learning setting characterised by the externalising of internal capabilities.

If we are to assume that business relationships function according to the empirical results presented by those concerned with industrial marketing, the executive committee does have sufficient knowledge about their most important customers and suppliers. That is, they interact with their immediate environment, an environment in which the actors involved take an active part in the process of mutual deployment of resources, learning from one another when combining knowledge accumulated apart. Consequently, under this assumption, the executive committee does have a rather comprehensive knowledge about the environment surrounding the organization to which they belong.

And let us then add one of the basic findings contributed by accounting scholars, i.e., that actions receiving attention in the form of measurement and rewards also will affect actions to come. The combination of these empirical findings makes it possible to assume that an executive committee will influence their organization's business relationships by designing and using a BSC. Simultaneously, as was the case with ABB and Skandia, the executive committee is not only aware of the well-being of their most important customers, but it also works closely with them in various projects. Hypothetically, the executive committee and the customer's executives are then in a position to control to some extent the means and ends of the interorganizational learning function. This can be achieved if the executives in each firm direct their attention to specific aspects in their respective operations, thereby increasing interorganizational learning in prioritised areas that would not otherwise receive as much attention. Of course, interorganizational learning will always occur in areas parallel to those prioritised. Nevertheless, it seems reasonable to believe that interorganizational learning in prioritised areas will benefit at the expense of those not prioritised, thus consuming effort and time that otherwise would be spent elsewhere. This possibility – executives together controlling at least some parts of the learning in a focal relationship – only exists within a limited period of time. In the long run, however, the dynamics of the industrial network, shaped over time by entries and exits, will exert its influence on the focal relationship in the form of changed demands on the process of mutual exchange between them. Still, the discussion indicates that executives do have some means of controlling how the immediate environment affects the organization to which they belong.

This paper has shown that learning in the form and shape of Kaplan and Norton's BSC has an internal focus. That is, it does not fully embrace the learning taking place in long-lasting exchange processes in business relationships. Thus, depending on what scenario management chooses to use, certain interorganizational learning implications will necessarily follow. Trying to control learning, not only vertically but also horizontally, may enforce learning in the predefined area, but it will exclude other business opportunities that would be possible to render through ongoing exchange processes. However, if one is to present more accurate conclusions, more comprehensive empirical examinations concerning the BSC's design and use in business relationships are required. In particular, the results from this study indicate a need for in-depth investigation of firms positioned in industrial markets, i.e. in industrial networks, and in which the BSC has been used for a period of time. Therefore a promising area for future research would be the role of accounting techniques in the dynamics of industrial networks.

References

ANDERSON, JAMES C., HÅKANSSON, HÅKAN, and JOHANSON, JAN, 1994, Dyadic Business Relationships Within a Business Network Context *Journal of Marketing*, Vol. 58, October, pp. 1–15.

BENGTSSON, LARS, 1997, Processutveckling med bas i produktionsgrupper – En studie av ABB Control AB i perspektivet affärsutveckling genom integration av Människa – Teknik – Organisation (MTO) Working paper. Gävle: University College of Gävle-Sandviken.

EWING, PER, 1995, The Balanced Scorecard at ABB Sweden, Stockholm: EFI Research Paper 6554. Stockholm School of Economics.

EWING, PER, 1996, Balanced Scorecards in use, Working paper. Stockholm: Stockholm School of Economics.

EWING, PER, and LUNDAHL, LENNART, 1996, The Balanced Scorecards at ABB Sweden – the EVITA projects, Stockholm: EFI Research Paper 6567. Stockholm School of Economics.

GRÖNLUND, ANDERS, 1989, *Lokal ekonomi – En fältstudie från tre produktionsavdelningar vid Volvo Komponenter AB*, Göteborg: BAS.

HÅKANSSON, HÅKAN, 1989, *Corporate Technological Behavior. Co-operation and Networks*, London: Routledge.

HÅKANSSON, HÅKAN, 1990, Technological Collaboration in Industrial Networks *European Management Journal*, Vol. 8, No. 3, September, pp. 371–379.

HÅKANSSON, HÅKAN and JOHANSON, JAN, 1993, The Network as a Governance Structure – Interfirm Cooperation Beyond Markets and Hierarchies in Grabher, G. (ed.), *The Embedded Firm. The Socio-Economics of Industrial Networks*, pp. 35–51. London: Routledge.

HÅKANSSON, HÅKAN and SNEHOTA, IVAN, 1989, No Business is an Island *Scandinavian Journal of Management*, Vol. 5, No. 3, pp. 187–200.

HÅKANSSON, HÅKAN and SNEHOTA, IVAN (eds.), 1995, *Developing Relationships in Business Networks*, London: Thomson Business Press.

HALLÉN, LARS, JOHANSON, JAN and SEYED-MOHAMED, NAZEEM, 1991, Interfirm Adaptation in Business Relationships *Journal of Marketing*, Vol. 55, April, pp. 29–37.

HOFFECKER, JOHN and GOLDENBERG CHARLES, 1994, Using the Balanced Scorecard to Develop Companywide Performance Measures *Journal of Cost Management for the Manufacturing Industry*, Fall, pp. 5–17.

JÖNSSON, STEN, 1996, *Accounting for Improvement*, Oxford: Pergamon.

JÖNSSON, STEN and GRÖNLUND, ANDERS, 1988, Life with a sub-contractor: New technology and management accounting *Accounting, Organizations and Society*, Vol. 13, No. 5, pp. 512–532.

KAPLAN, ROBERT S. and NORTON, DAVID P., 1992, The Balanced Scorecard – Measures That Drive Performance *Harvard Business Review*, Jan.-Feb., pp. 71–79.

KAPLAN, ROBERT S. and NORTON, DAVID P., 1993, Putting the Balanced Scorecard to Work *Harvard Business Review*, Sept.-Oct., pp. 134–147.

KAPLAN, ROBERT S. and NORTON, DAVID P., 1996a, *The Balanced Scorecard – Translating strategy into action*, Boston: Harvard Business School Press.

KAPLAN, ROBERT S. and NORTON, DAVID P., 1996b, Using the Balanced Scorecard as a Strategic Management System *Harvard Business Review*, Jan.-Feb., pp. 75–85.

LUNDAHL, LENNART and EWING, PER, 1997, ABB's EVITA Puts Customer-Focused Control Centre Stage *Measuring Business Excellence*, Vol. 1, No. 3, pp. 24–29.

MAISEL, L. S., 1992, Performance Measurement: The Balanced Scorecard Approach *Journal of Cost Management for the Manufacturing Industry*, Summer, pp. 44–49.

MOURITSEN, JAN, HØHOLDT, JAKOB and JØRGENSEN, ANDERS A. V., 1995/96, De »nye« og de »gamle« ikke-finansielle nøgletal *Økonomistyring & Informatik*, 11. årgång, Nr. 6, pp. 387–409.

OLVE, NILS-GÖRAN, ROY, J. and WETTER, M., 1997, *Balanced Scorecard i svensk praktik*, Stockholm: Liber Ekonomi.

SIMONS, ROBERT, 1990, The Role of Management Control Systems in Creating Competitive Advantage: New Perspectives *Accounting, Organization and Society*, Vol. 15, No. 1/2, pp. 127–143.

SIMONS, ROBERT, 1991, Strategic Orientation and Top Management Attention to Control Systems, *Strategic Management Journal*, Vol. 12, pp. 49–62.

SIMONS, ROBERT, 1995, *Levers of Control – How Managers Use Innovative Control Systems to Drive Strategic Renewal*, Boston: Harvard Business School Press.

SOLLI, ROLF, 1991, *Ekonomi för dem som gör något: en studie av användning och utformning av ekonomisk information för lokala enheter*, Lund: Studentlitteratur.

WENNBERG, INGE, 1994, På väg bort från ekonomistyrningen, *Ekonomi & Styrning*, Nr. 2, pp. 6–10.

WESTIN, OLLE, 1993, *Informationsstöd för lokal ekonomi – En studie kring centrala informationssystem och lokala informationsstöd ur ett verksamhetsperspektiv*, Göteborg: BAS.

CHAPTER 4

Business-Governed Product Development: Knowledge Utilization in Business Relationships

ULF ANDERSSON and JONAS DAHLQVIST

Introduction

A changing world economy forces today's business enterprises, more than ever, to be keenly alive to shifts in market preferences. Market vicinity, organizational learning, and shortened product development cycles, to name but a few critical success factors, are the focus for the manager of the 90s. Increased availability of advanced production technology has made cost cutting moves easily imitated, while ingenious application of technology in product development remains a difficult-to-manage activity, and thus difficult to imitate and replicate. Transfer of knowledge through the use of Global Best Practices and Benchmarking seems to lend itself better to cost-cutting refinements of processes dealing with known and stable goals than to revenue-increasing interventions in unstructured innovation processes. The counterintuitive effect of these observations is that sustainable competitive advantage is gained not by reducing costs, but by increasing revenues. The reason is, of course, that in the long run no competitive advantage is gained by implementing solutions that competitors would most likely be able to imitate in the near future. The slogan of the Boss fragrance advertisement – "Don't imitate, innovate!" – highlights a true challenge for anyone committed to product development. This paper takes up the gauntlet, and hopefully adds to our knowledge about product development by posing the question:

How can knowledge located at different places within an economy be utilized through business exchange in order to enhance product development?

A well-articulated theory of the business enterprise, and of the role played by knowledge in business activities, are indispensable prerequisites for a fruitful discussion of product development. The rationale for this assumption is simple: sustainable competitive advantage gained by product development is inherently dependent on the commercial viability of new products introduced to the market, and this, in turn, depends on a true understanding of the conditions surrounding the business activity. These conditions will be more and more associated with knowledge management within the business place (Ottum and Moore 1997). The paper departs from the so-called markets-as-networks perspective – a theoretical perspective concerned with business exchange in industrial markets – which emerged in Sweden during the 1980s. The concept of "business enterprise" is constructed from this perspective (Snehota 1990). Thereafter the role played by knowledge in business exchange is discussed. The outcome of business-governed product development is dependent on the possibility of making use of knowledge located at different areas in the economy. The mobility of business knowledge is therefore discussed with reference to a concept of knowledge influenced by cognitive science (Blackler 1995). The idea of a business relationship is introduced as one among many ways in which business activity could be conducted. Research within the markets-as-networks perspective has shown that long-lasting business relationships are effective mechanisms for coordinating knowledge-intensive exchange situations. Research has further shown that long-lasting business relationships exhibit a high level of product development.

The arguments elaborated in the paper are synthesized in a discussion of the business relationship as mediator of knowledge in business-governed product development. Propositions concerning the relation between relationship characteristics and product development are presented, followed by an empirical test. The paper ends with a discussion of the results, and priority is given to managerial implications.

Markets-as-networks and business enterprises

One of the most salient results of empirical research carried out within the markets-as-networks perspective concerns the observation that business exchange on industrial markets is conducted in a network of long-lasting relationships (Hallén 1986). In its most general form, the explanation as to why business is conducted within long-lasting relationships states that in business situations characterized by changing and strong interdependencies between business actors the business relationship is a more effective device for coordination of resources than is the market mechanism (Håkansson and Johanson 1993). Business enterprises need to be exchange effective as well as production efficient, and in business situations such as the one described, management of exchange interdependencies takes precedence over management of internal production processes, resulting in closer contact between supplier and customer. In short, business exchange should be conceptualized as an organization process where knowledge about customers' needs and knowledge about production capabilities is moulded into new business opportunities.

Counterparts' needs are found in the market, whereas production capabilities are located inside the business enterprise, which means that an important function in business exchange concerns the simultaneous management of knowledge about these needs and knowledge about production capabilities. As knowledge often turns out to be difficult to transfer (*cf.* Blackler 1995; Madhok 1997; Malmberg, Sölvell, and Zander 1996) we will conceive of the functionary (or rather the function) in charge of this process as a *knowledge mediatior* rather than a *knowledge manager*. Figure 4.1 illustrates the role played by the knowledge mediator in business exchange.

FIGURE 4.1
The knowledge mediator moulding counterparts' needs and production capabilities

In Figure 4.1, the knowledge mediator moulds knowledge about counterparts' needs and knowledge about production capabilities into new business opportunities. During this process the supplier develops new knowledge concerning how-to-produce products useful for potential customers, and the customer develops new knowledge concerning how-to-use products supplied by the market. This simultaneous development of knowledge concerning how-to-produce and how-to-use constitutes the substructure of the process through which business actors sense new business opportunities, i.e. understand how they could gain from exchange with each other.

In business activity, knowledge about how-to-produce and knowledge about how-to-use is not confined to knowledge about technical performance of production technology and products. Equally important is the ingenuity of the actor in finding problems and possibilities to which to apply the technology and the products. The development of that knowledge is to a large extent a social activity; problems and possibilities never exist *per se* but rather come into existence during day-to-day activities and in relation to goals aimed for by the actors.

Socially constructed knowledge, such as knowledge about how-to-use and how-to-produce, is "sticky" (it tends to stick to the social system in which it is constructed), and the possibility of our decontextualizing it and transferring it through time and space is limited (Malmberg, Sölvell and Zander 1996). Blackler (1995) noted four characteristics of such knowledge. First that knowledge is *situated,* meaning that often the relevant knowledge is too specific to be named and assessed but exists only as specific knowledge in specific situations. Secondly, that knowledge is *pragmatic*; thus knowledge and cognition are not only the result of internal manipulation of ideas, but also the result of physical, manual, and interactional actions. Thirdly, that knowledge is *provisional* and as it is continuously evolving its validity is frequently limited to the situation in which it is used. Fourthly, that knowledge is *mediated,* that is, rather than being transmitted, knowledge is carried and manipulated by actors engaged in interaction with other actors as well as with artifacts. Blackler concluded that these characteristics imply that knowledge should be analyzed as an active process, as knowing, where the focus should be on the activity system through which people achieve their knowing.

Even though some knowledge could, of course, be decontextualized and transferred, studies within the markets-as-networks perspective support the view that important aspects of business knowledge relevant for product development are situated, pragmatic, and provisional, and thus mediated through people's actions and continuous communication, rather than being transferred. The possibility of managing knowledge necessary for product development at a distance is thus limited. In product development processes, organizational learning, to use a popular buzzword, demands not only proximity of a market, but also what might be called counterpart vicinity within activity systems. Activity systems are, to a large extent, constituted of individuals with diverging interests and priorities, which implies that incoherence, paradoxes, and conflicts accompany the mediation of knowledge. Tensions in activity systems are inevitable, but treated correctly, they could provide a potential driving force for change. As noted by Blackler, "new ways of knowing and doing can emerge if communities begin to rethink what, in a different context, Unger (1987) called the 'false necessity' of everyday life, and to engage with the tensions in their activity systems" (Blackler 1995: 1038).

Relationship characteristics and product development

The business relationship is exchange effective in knowledge-intensive business situations that are characterized by many and changing interdependencies between the business actors (Håkansson and Johanson 1993). The reason hereto is that the relationship enables the actors to fairly well access both the production system in which the counterpart operates and the cognitive model of the system held by the counterpart. The business relationship could be said to connect different resource and activity structures on a subjective or cognitive level and thus facilitate mutual learning concerning interdependencies between customers and suppliers.

Figure 4.2 shows how three interrelated processes impact on the supplier's ability to learn about the customer and to develop new and commercially viable products. The most obvious process is indicated by arrow A, where the supplier relates its own how-to-produce knowledge to internal activities and resources. Besides this internal focus the supplier is able to evaluate its how-to-produce knowledge in the light of the needs of its customer. Customer needs are evaluated through two processes. First, they are evaluated through examination of the customer's how-to-use knowledge, that is through examination of the customer's own view of problems and possibilities it faces (arrow B). Secondly, they are evaluated through examination of problems and possibilities faced by the customer according to the supplier (arrow C). Phrased differently, the business relationship makes it possible for the supplier to reason with the customer about its needs and its view of why it has these needs. The relationship also enables the supplier to independently assess the situation faced by the customer. That is, through close discussions and field visits, the relationship enhances the possibility of connecting

FIGURE 4.2
The relationship as knowledge mediator in product development

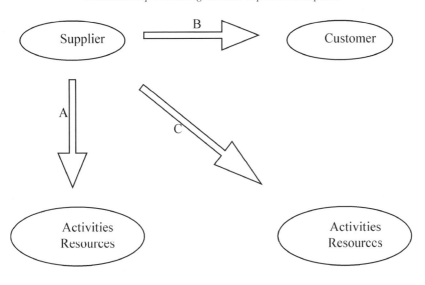

people engaged in activities that may be important mediators of knowledge relevant for product development.

In the next section three propositions concerning the relation between characteristics of the relationship and product development will be presented. The guiding question during this discussion has been *What kinds of characteristics of the business relationship could be expected to impact on product development?* Since one of the core arguments of the paper is that the relevant knowledge may be difficult to map and label, and that we should therefore focus on the activity system through which people achieve their knowing (Blackler 1995), the characteristics that constitute such a system will be highlighted.

Three relationship characteristics; *interaction intensity, easily managed adaptations,* and *relationship duration*

As pointed out by Blackler, relevant knowledge about how-to-produce and how-to-use appears to be sticky, and rather than being transferred in channels, it is mediated through activity systems. The business relationship has been conceptualized as one such activity system, and the counterpart vicinity enabled by the relationship has been pointed out as one of the general characteristics that renders the use of knowledge in product development more effective. From the discussion it follows that the intensity of the interaction, in a general sense, will impact on the ability of the actors involved in the relationship to successfully carry through a development project. Here three dimensions of the characteristic *interaction intensity* will be discussed, namely, number of people involved in the relationship, number of different knowledge areas involved in the relationship, and frequency of interaction within the relationship.

In business-governed product development, the actors involved are part of a complex activity system embracing knowledge and capabilities located at many different places. As knowledge is situated – which means that as a result of specificity it is difficult to identify, assess, and name – product development is dependent on incremental sense-making and on incremental trouble-shooting (Malmberg, Sölvell and Zander 1996) in situations where actors can access each other easily. Many individuals are involved in the business activity, and each of them may be knowledgeable about something that could turn out to be useful for development of the product. A high level of close interaction characterizes the business relationship. It could, therefore, be expected that the relationship would be positively correlated with product development. If knowledge had been general and abstract, individual involvement would not have been as crucial, as knowledge could then have been mapped and transferred. In other words we would expect that involving more people in the business relationship would increase the probability of successful product development. "People" should here, as elsewhere in the paper, be understood as the people normally engaged in the actual business activity in one way or another. There is thus a natural upper limit to the number of people that could be engaged in the business relationship, and it is of course not suggested that effectiveness in product development could easily be improved by staffing the relationship, or development project, with an excessively large number of people.

Knowledge is systematically distributed within a business exchange system. A customer may be more knowledgeable than a supplier may be concerning how to use the product sold by the supplier, whereas the supplier knows more about how to produce the product. As the business enterprise is not a coherent or monolithic subject, but rather an organizational system constituted of individuals and departments, we would expect to find a division of knowing within as well as between business enterprises. Knowledge about how-to-use and knowledge about how-to-produce ought to be systematically distributed within business enterprises: marketing knows more about how-to-use than production does, whereas the latter knows more about how-to-produce. A business actor should not be seen as either a user or a producer of products. Even though a focal supplier is a producer of products supplied to the market, it is a user of products bought from third party suppliers. Although our focal production department knows how to use the products supplied by third parties, they may be less knowledgeable about the production of them, and thus less knowledgeable about the possibility of modifying them in order to be better equipped to serve customers' needs. Procurement departments, with their relationship to suppliers, could be expected to be in a better position than marketing and production departments to judge the possibilities of modifying production equipment. To the extent that creative application of production equipment is dependent on this knowledge, it would thus be desirable to involve procurement in the development process. The involvement of individuals from different knowledge areas ensures that different perspectives are applied to the product and the development process. Involvement of different knowledge areas elicits and questions the false necessity of everyday life, as access to different forms of knowledge will be secured and new ways of knowing and doing will be produced. As business relationships are often task-driven, they could be expected to include individuals from different knowledge areas, and to be positively correlated with product development. In other words we would expect involvement of different knowledge areas in the development process to increase the probability of success in product development.

Knowledge relevant for business-governed product development is only to a minor extent abstract and generic, and is continuously evolving during the acting and interaction of the actors involved. Knowledge is, thus, provisional. Depending on the purpose of the business relationship – the character of the development project – different phases of the transformation of the knowledge involved will be more or less useful for the project. As a result it is difficult to plan when useful knowledge will be accessible, and it is also difficult to store once it has come into existence. In view of the difficulty in managing knowledge, managers in charge of product development should rather focus on organization of the knowledgeable individuals involved. It becomes important to have "the right man at the right place" as knowledge evolves. This is a critical aspect of product development, since it is impossible to say exactly where to place whom at what time, and it does not suffice to put just anyone, anywhere, at any time. This will be a predominant concern during the development of knowledge strategies launched by firms today in their efforts to establish knowledge management programs. Although it is difficult to determine who should be involved at what point in a development project, it could be expected that a business relationship, with its documented high frequency of interaction, would be positively correlated with product development.

The number of people involved, involvement of different areas of knowledge, and frequency of interaction, are the three dimensions that constitute the variable interaction intensity. Intensity of the interaction is expected to be positively correlated with successful product development.

Proposition 1: The higher the intensity of the interaction in a business relationship, the higher the probability of successful product development.

If it is accepted that knowledge relevant for business-governed product development is not abstract and generic, it follows that development and utilization of knowledge in development projects are not the results of mere internal manipulation of ideas later transferred to the appropriate user. Knowledge is pragmatic – it is the result of physical, manual, and interactional actions as well as of pure manipulation of ideas (Blackler 1995). This implies that knowledge evolves as collective action is undertaken. When people work together knowledge evolves in a situation where the actors involved are fairly well able to communicate and demonstrate their own understanding of the task to be accomplished, and to appreciate the conception held by other actors involved in the relationship. This would imply that easily managed adaptations, more dependent on allocation of motivation than on development of capabilities, could function as effective mediators of knowing between actors involved. As the actors involved start to work together, the rationale for each actor's behaviour surfaces. These rationales may function as a frame of reference for future interaction in the development process. Even if incoherence and tensions in a relationship could be confronted at an intellectual level, the situated and pragmatic nature of knowledge implies that the full effect of such a confrontation, and thus the most effective way to overcome the false necessity of everyday life, is achieved when the implications of a clash between conflicting ideas are put into practice (Tabrizi and Walleigh 1997). We would expect the business relationship with its task oriented and co-operative character to be positively correlated with product development. More precisely:

Proposition 2: The existence of easily managed adaptations between actors in a relationship will increase the probability of successful product development.

A large extent of the knowledge involved in business-governed product development is the result of social interaction. Knowledge about how-to-use has developed as business actors have assessed the business situation they face: identified problems and possibilities in this situation; worked out solutions, which they have tried to picture to others as sensible; and finally tried to produce and sell. Knowledge about how-to-use is thus influenced by socially constructed realities. In business-governed product development, where customers and suppliers are connected, different social realities are connected, which means that different ways of viewing common activities – product development – are confronted. These realities could be seen as interpretative schemes of a highly tacit nature, which become understood by an outsider through lengthy reasoning with the actors embraced by (or embracing) a particular interpretative scheme. Taken-for-granted conceptions about how things should be done, and what might be done, are often unconsciously held by the actors, and thus are not easily communicated. To the extent that taken-for-granted conceptions – the false necessity of everyday life – impacts on product development, the duration of the business relationship ought to be correlated with product development. It takes time for the actors involved to come to grips with the way others define problems and interpret possibilities and solutions. Since business relationships are often long-lasting they could be expected to be positively correlated with product development. More precisely:

Proposition 3: The longer the duration of a business relationship the greater the probability of successful product development.

Empirical analysis

In this section the variables needed for testing the propositions described above will be operationalized. The propositions will be tested by formulating a regression model with "successful product development" as the dependent variable and "interaction intensity", "easily managed adaptations" and "relationship duration" as independent variables.

Ulf Andersson and Jonas Dahlqvist

The sample used consisted of 277 customer relationships with 97 suppliers. The original study aimed at investigating the network relationships of subsidiaries to Swedish multinational corporations. In that study, called Managing International Networks, personal interviews were undertaken with the CEO, the manager responsible for purchasing, and the sales manager. They were asked to choose three of their most important relationships in each of the categories: supplier, customer, and other. Using an extensive standardized questionnaire, they were then interviewed about each of the selected relationships. The data about customer relationships, which were gathered in the original study, have been used in this paper. This makes it possible to view the subsidiary as a supplier and to study its most important customer relationships. For a further description of the original study see Andersson (1997) and Pahlberg (1996).

Indicators of interaction intensity

Three indicators were used to assess the interaction intensity in the supplier–customer relationships. The respondents were asked to estimate how many people were involved in the relationship from the supplier's side. The number of people involved was translated into a scale ranging from *1 = a few*, to *3 = many*. They also indicated the frequency of direct contacts between the supplier and the customer. The frequency of direct contacts was translated into a scale ranging from *1 = less than six times a year*, to *5 = once a week or more often*. These two indicators, together with the different knowledge areas engaged in direct contacts with the customer from the supplier's side, made up the variable of interaction intensity. The indicator "different knowledge areas" corresponded to purchasing, production, marketing and sales, R&D, top management, and administration. The more of these areas involved in direct contacts the higher the score, which meant that the scale ranged from 1 to 6. The highest score, 6, appeared in thirteen cases and these were the only cases where administration was mentioned as an involved knowledge area in the direct contacts with the customer. The correlation matrix for this group of indicators shows relatively high interitem correlation: see Table 4.1. Although the indicators do not have a very high reliability coefficient (Cronbach's α = 0.56) some overlap between the indicators could be identified. Therefore, we decided to combine the indicators into one measure by using principal component analysis (PCA). The factor scores of the first principal component were used to represent the intensity in the relationship between the supplier and the customer. By using PCA we receive a standardized variable with the mean 0 and the standard deviation 1, where the indicators are weighted.

62

TABLE 4.1

Correlation matrix for the indicators of interaction intensity

	Number of people	Number of different knowledge areas
Number of different knowledge areas	0.57 p=0.00	
Frequency of direct contacts	0.37 p=0.00	0.23 p=0.00

Indicators of easily managed adaptations

Easily managed adaptations are adaptations that are limited by the supplier's motivation rather than its capabilities. The supplier in an exchange relationship can adapt itself in several dimensions in order to show its commitment to the customer. In this paper, however, the motivational aspect of adaptations is not in focus, but rather the cognitive aspect. When people work together knowledge evolves in a situation where the actors involved are fairly well able to communicate and demonstrate their own understanding of the task to be accomplished, and to appreciate the conception held by other actors involved in the relationship. As the actors involved start to work together, the rationale for each actor's behaviour surfaces. These rationales may function as a frame of reference for future interaction in the development process.

The suppliers were asked to estimate their adaptation to each particular customer in terms of four dimensions, namely, business conduct, organization structure, production technology, and product technology. The correlation matrix for this group of indicators (see Table 2) shows high interitem correlation. The indicators also have a relatively high reliability coefficient (Cronbach's $\alpha = 0.76$) indicating an overlap between the indicators. Easily managed adaptations work as cognitive co-ordinators of the actors' perceptions of the co-operation. Principal component analysis was again used to combine the indicators into one measure and receive a standardized variable. The factor scores of the first principal component were used to represent the easily managed adaptations made by the supplier.

In order to avoid problems with causality between the independent variable "easily managed adaptations" and the dependent variable "successful product development" we also constructed a variable called "easily managed adaptations[i]" in which the indicator "adaptation of product technology" was omitted. As can be seen in the correlation matrix, see Table 4.2, the interitem correlation is slightly lower if the indicator "adaptation of product technology" is omitted. The reliability coefficient decreases somewhat (Cronbach's $\alpha = 0.66$) but still indicates an overlap between the indicators.

TABLE 4.2
Correlation matrix for the indicators of easily managed adaptations

	Adaptation of business conduct	Adaptation of organization structure	Adaptation of production technology
Adaptation of organization structure	0.47 p=0.00		
Adaptation of production technology	0.32 p=0.00	0.36 p=0.00	
Adaptation of product technology	0.49 p=0.00	0.33 p=0.00	0.59 p=0.00

The indicator of relationship duration

Together with the independent variables "easily managed adaptations" and "interaction intensity" it could be assumed that the duration of the relationship between the partners in the business relationship should be positively related to successful product development. Over time, the supplier and customer learn how to do business and interact with each other. As a proxy for duration of the relationship the age of the business relationship will be used. The respondents indicated the supplier's first contact with the customer for each one of the relationships. The mean age of the studied relationships was high at 18.3 years, and only 25 relationships (9%) had existed for less than three years. Based on this low number, it is difficult to detect the impact of newly established relationships.

The indicator of successful product development

Successful product development focuses on the extent to which the customer's knowledge about how-to-use can improve the supplier's product development function. Successful product development can be estimated *ex post facto* by looking, for example, at sales volume increase. This would require two measures separated in time. Since the interviews were made at one point in time another type of measurement had to be used, namely, the respondents' perception of customer importance for the supplier's product development. Using a five-point Likert scale the sales managers for each supplier estimated the importance of each of the customers for its product development.

Relationship impact on successful product development

In this section the operationalized variables will be used in two linear regression equations, shown in Table 4.3. The reason for using two regression equations is the possible problem of causality between the indicator "adaptation of product technology" and the dependent variable "successful product development".

TABLE 4.3
The regression equations of product development

Dependent variable	Independent variables Estimated coefficients, T-ratios (in parenthesis) and significant levels					
Successful product development	Easily managed adaptations	Easily managed adaptations[i]	Interaction intensity	Relationship age	R^2	F-value
(I)	0.79 (12.87)‡		0.15 (2.39)*	0.01 (1.52)	0.42	63.61‡
(II)		0.55 (7.55)‡	0.16 (2.23)*	-0.001 (-0.03)	0.22	25.24‡
*p < 0.05; †p < 0.01; ‡p < 0.001						

The main focus of this paper is not to build a fully-fledged model of product development, but rather to investigate some typical features of the business relationship – easily managed adaptations, interaction intensity, and relationship age – and their impact on successful product development.

The equations in Table 4.3 are basically the same except for different operationalization of the independent variables "easily managed adaptations" and "easily managed adaptations[i]". The results displayed in Table 4.3 demonstrate that "interaction intensity" significantly explains the variation in the dependent variable "successful product development" (*t*-values 2.39 and 2.23; $p < 0.05$). This result supports the first proposition that: "The higher the intensity of the interaction in a business relationship, the higher the probability for successful product development." The regression equations also show that "easily managed adaptations" and "easily managed adaptations[i]", are both highly significant (*t*-values = 12.87 and 7.55, $p < 0.001$) in explaining the variation in the dependent variable. This gives support to proposition 2, which stated that easily managed adaptations are positively correlated with successful product development.

The duration of the business relationships did not seem to have any significant impact on successful product development for the supplier.

The regression equations above do not support proposition 3. Although this result is somewhat surprising, one explanation could be the relatively high age of the business relationships in the sample. Another explanation could be the lack of variation in this independent variable.

Implications for practice

Sustainable competitive advantage is gained through ingenious use of a firm's knowledge for revenue enhancement rather than for cost containment. Product development, most likely the single most important factor for increased revenue generation, is not managed separately from a firm's day-to-day business activity, since commercial viability of new products demands alignment with customer preferences. Market knowledge has to be gathered and used together with production knowledge. The arguments elaborated in this paper boil down to two general insights. First, knowledge needed for successful product development may turn out to be difficult to move in time as well as in space due to social embeddedness. Secondly, effective utilization of knowledge located at different places *could* be achieved if the nature of business knowledge is understood and utilization is seen as an organizational rather than analytical issue. This means that, rather than trying to map, store, and distribute knowledge, managers should focus on the establishment of effective task organization.

In the paper, strong support was found for the two propositions claiming that interaction intensity and existence of easily managed adaptations are positively correlated with successful product development. Evaluated in the light of today's focus on information technology (IT) as the key enabler for effective management of organizational knowledge, the results of this study imply that managers should be cautious about using IT too aggressively in the quest to speed up knowledge management and product development, at the expense of personal involvement. Even if IT provides excellent opportunities for sharing and storing of knowledge, important aspects of knowledge necessary for business development may escape key actors if IT is not supported by an appropriate organizational structure and interactional mode.

An appropriate organizational structure refers to the involvement of people from different knowledge areas within a firm. Once again the reader should be reminded that the discussion concerns the normal population of a business relationship. Also important is the involvement of people from the customer organization. An appropriate interaction mode implies face-to-face contact, field visits, and, when possible, adaptations in product and production technology. Appropriate organization structures and interaction modes will secure an activity system through which business knowledge located in different places could be mediated to the individuals most capable of using it effectively.

As demonstrated by research conducted within the markets-as-networks perspective, such activity systems, in the form of business relationships, seem to develop spontaneously. The results from this paper help business managers better perceive that these kinds of activity systems are of great importance for the firm's continuous development. The response to this insight may range from rather passive caretaking of existing business relationships, to more proactive development of strategic business relationships.

Today, many firms have realized the need to take care of what is commonly labelled their "intellectual capital". It is assumed that a mandatory step in this process is to set up a knowledge strategy to guide the firm's knowledge management process. It is further assumed that the knowledge strategy has to be aligned with the firm's general business strategy. This may turn out to be easier said than done, due to the difficulty of conceptualizing knowledge relevant for the business strategy. This paper suggests that instead of focusing directly on knowledge necessary for a firm's continuous development, managers ought to depart from existing business relationships and evaluate their relative strategic importance. This results in a more tangible design of the strategy process compared with a pure and decontextualized knowledge approach. More specifically, business managers ought to evaluate existing business relationships from the point of view of the business strategy. The focus during this evaluation process should be on the relationships that they think could improve the firm's knowledge and developmental capabilities. Identified business relationships should be given priorities when it comes to allocation of resources that may improve interaction intensity and the existence of easily managed adaptations.

Workshop techniques have been developed to assist in bringing mental models held by actors to the surface (Senge, Kleiner, Roberts, Ross, and Smith 1994). Such techniques offer a promising future in facilitating the work of cultivating business relationships that effectively mediate business knowledge. This is so even though it has been argued that true leverage of confronted incoherence and tensions in a business relationship happens when an interaction moves from the intellectual to the practical level. Social worlds brought into a business relationship by actors contain powerful mental models that, in a very real sense, impact on the ability of the actors to co-operate in a creative manner. Awareness of the existence of such mental models, and the possibility of bringing them to the surface and reflecting upon them, may render the business relationship more effective as a mediator of business knowledge.

To sum up, the paper highlights the role played by existing business relationships in a firm's continuous business development. Further, the paper demonstrates the need to pay close attention to a firm's strategic business relationships and to manage them in terms of interaction intensity and operational co-operation, in the paper termed "easily managed adaptations". Such management may range from passive monitoring and caretaking, to more proactive relationship development. It has been suggested that relationships should be evaluated from, and linked to, a firm's business strategy, and supported by facilitation techniques where mental models held by actors are brought to the surface and confronted.

References

ANDERSSON, U., 1997, *Subsidiary Network Embeddedness: Integration, Control and Influence in the Multinational Corporation*, Published Doctoral Diss., Dpt. of Business Studies, Uppsala University.

BLACKLER, F., 1995, Knowledge, Knowledge Work and Organizations: An Overview and Interpretation *Organization Studies*, vol. 16, pp. 1021–1046.

HÅKANSSON, H. and JOHANSON, J., 1993, The Network as a Governance Structure: Interfirm Cooperation Beyond Markets and Hierarchies, in Grabher, G. (ed.), *The Embedded Firm*, London: Routledge, pp. 35–51.

HALLÉN, L., 1986, A Comparison of Strategic Marketing Approaches, in Turnbull, P. and Valla, J.-P. (eds), *Strategies for International Industrial Marketing*, London: Croom Helm.

MADHOK, A., 1997, Cost, Value and Foreign Market Entry Mode: The Transaction and the Firm *Strategic Management Journal*, Vol. 18, No. 1, pp. 39–61.

MALMBERG, A., SÖLVELL, Ö. and ZANDER, I., 1996, Spatial Clustering, Local Accumulation of Knowledge and Firm Competitiveness *Geografiska Annaler*, 78 B (2), pp. 85–97.

OTTUM, B. D. and MOORE, W. L., 1997, The Role of Market Information in New Product Success/Failure *Journal of Product Innovation and Management*, vol. 14, pp. 258–273.

PAHLBERG, C., 1996, *Subsidiary – Headquarters Relationships in International Business Networks*, Published Doctoral Diss., Dpt. of Business Studies, Uppsala University.

SENGE, P., KLEINER, A., ROBERTS, C., ROSS, R. and SMITH, B., 1994, *The Fifth Dimension – Fieldbook*, New York: Currency Doubleday.

SNEHOTA, I., 1990, *Notes on a Theory of Business Enterprise*, Uppsala University, Department of Business Studies (diss.).

TABRIZI, B., and WALLEIGH, R., 1997, Defining Next-Generation Products: An Inside Look *Harvard Business Review*, November-December, pp. 116–124.

UNGER, R., 1987, *False necessity; Anti-necessitarian social theory in the service of radical democracy*, Cambridge: Cambridge University Press.

CHAPTER 5

The Internationalisation Process as Knowledge Translation in International Business Relationships

SOON-GWON CHOI and KENT ERIKSSON[1]

Introduction

Process models have played an important role in research on the internationalisation of firms (Andersen 1993; Bilkey and Tesar 1977, Carlson 1974, Cavusgil 1984, Czinkota 1982, Johanson and Vahlne 1977; Reid 1983). The Johanson and Vahlne (1977) model identifies an incremental interplay between market knowledge and commitment to do business in the foreign marketplace. But most of these studies have not explicitly modelled how to investigate the internationalisation process empirically. Instead, most studies have focused on the internationalisation behaviour as a consequence of the process models. The aim of this paper is to explicitly model an empirical research framework for the internationalisation process.

Toyne (1989) proposes that international exchange should be the foundation for theory building in international business. We propose that international exchange is framed within international business relationships, and consequently view international business relationships as the foundation for theory building in international business. Business relationships have been found to be long term (Blois, 1972; Ford, 1990) and

---

[1]The authors appear in alphabetical order and have contributed equally to this paper.

involve considerable investments by parties as they make adaptations to each other (Hallén, Johanson, and Seyed-Mohamed 1991). The exchange within a business relationship usually ties the parties together in a wider sense since it leads to an increase in understanding, trust, and commitment between the parties (Alter and Hage 1993; Axelrod 1984). The business relationship thus becomes the frame within which parties learn about each other.

But the understanding of what parties learn about each other in international business relationships is not very clear. To accomplish the aim of explicitly modelling an empirical research framework for the internationalisation process we need to develop an understanding of what is learnt in international business relationships. Learning can be conceptualised as a knowledge transfer process. But the transfer of an object or idea implies that it is moved from one place to another relatively intact. This notion has been thought ill-suited for business research and it has instead been proposed to use the term translation (Latour 1986). Translation implies that a phenomenon in one context is moulded into another context. It is thus clearly not the same phenomenon in both contexts, rather it is similar in certain respects and different in others. Consequently, we define knowledge translation as the process where knowledge in one setting re-occurs in a modified form in another setting. Since knowledge translation in business relationships is the heart of the internationalisation process we state our purpose as follows: *The purpose of this paper is to explicitly model the process of knowledge translation between firms in international business relationships.* In order for the implications of this process to be explicit, we also include the outcome of this process in the form of gained benefits and costs incurred.

The paper is structured as follows. First, there is a review of the definition and classification of knowledge, and the process of knowledge acquisition. Second, there is a discussion on the extent of knowledge embeddedness in a local market. This embeddedness is the fundamental problem of the translatability of knowledge. And then, the process of knowledge translation and organisational learning process is discussed (Inkpen and Beamish, 1997). The result is a model of knowledge translation in international business.

The knowledge translation process

Research has shown that knowledge has a market or location specific character (Johanson and Vahlne, 1977; Kogut and Zander, 1993; Makino and Delios, 1997; Lyles and Salk, 1997). The embeddedness of knowledge to a particular setting is critical to determine its translatability. If the knowledge which is crucial to the firm is not market specific, then there is no problem in translating knowledge. The firm can use the knowledge as a public good (Kogut and Zander, 1993). But, as mentioned earlier, the knowledge that a firm has about doing business is tied in with the local setting. The embeddedness of knowledge is the fundamental issue for the translatability of knowledge. This section will develop a frame for understanding the embeddedness of knowledge. It will be done by first reviewing prior research relating to knowledge translation, then developing a view of how tacit and explicit knowledge affects translation, in order to arrive at the embeddedness of knowledge.

Previous research on knowledge translation

The critical issue in knowledge translation is that the knowledge should be of use and value in the context it is being translated into. This is exemplified by a study of Canadian retail firms entering the US (O'Grady and Lane 1996). The Canadian firms thought that the US market was similar to the Canadian, and translated their domestic business knowledge when expanding into the US. The expansions proved expensive and several firms retreated, whereas those who stayed moulded their business behaviour to the US setting. These findings show that translation into apparently similar markets was unsuccessful. Instead, business knowledge needed to be translated with great consideration for the foreign business context.

Hedlund (1994) suggested that firms can acquire, translate and transform knowledge by using several techniques, such as articulation of knowledge, a healthy dialogue between units and other firms, and dissemination of new knowledge. He proposes two dimensions for the analysis of knowledge translation: tacit and explicit. Tacit knowledge is nonverbalised or even nonverbalisable, intuitive, and unarticulated (cf. Polanyi 1962). Explicit knowledge is specified verbally, for example in writing, computer programs, patents, drawings. The difficulty of articulation is the criterion for the distinction between tacit and explicit knowledge.

Kogut and Zander (1993) extends Hedlund's research by studying how knowledge translation varies with the character of knowledge in the context of the multinational corporation. They proposed three dimensions for assessing tacitness and explicitness of knowledge: a) codifiability, which refers to the possibility to codify knowledge in scripts, such as documents, routines, or programs; b) teachability is the possibility of teaching others so that they acquire knowledge; and c) complexity is the difficult to understand knowledge. Tacit knowledge is difficult to codify and teach, and complex, whereas explicit knowledge is easy to codify and teach, and not complex. Their results show that as knowledge becomes more codified and more easily taught, the more likely it is to be translated to a third party rather than to a wholly owned subsidiary. They also found that as technologies increase in complexity, they are more likely to be translated to wholly owned subsidiaries rather than to a third party. Based on the argument that the integration with the firm is stronger for wholly owned subsidiaries than for third parties, these results show that knowledge that has a tacit character is more difficult to translate than more explicit knowledge.

Apart from Kogut and Zander, other researchers have found that knowledge can be translated across and within firms (Makino and Delios, 1997; Lyles and Salk, 1997). But effective translation requires benevolent conditions, such as frequent interaction between partners in business relationships. Interactions can involve, for instance, adaptations in production processes that makes firms learn about each other (Hamel 1991, Hamel, Doz and Prahalad, 1989: 134). Other examples that make organisational boundaries permeable are when firms have similar goals and interests (Buckley and Casson, 1976).

Makino and Delios (1997) focused on the organisational arrangements required for knowledge acquisition. They test the existence of three channels of local knowledge acquisition: by forming a joint venture with a local firm, by transfer from the foreign parent's stock of past host country experience, and by the accumulation of operational experience in the host country. They showed that each knowledge acquisition channel was found to influence performance. Together with the above results, Makino and Delios's show that knowledge can be transferred, but that it is not easy. Explicit knowledge can be transferred easily, but knowledge which is tacit can be transferred with difficulty. These findings point to that knowledge translation is possible.

Contrary to the above, Johanson and Vahlne (1977) argued in the Uppsala model that most business related knowledge is tacit, and thus not possible to translate. Barkema et al. (1996) found that firms' past experiences of the country where they are establishing is more important for learning about that market than experiences from other countries. Barkema et al. thus present strong support for the Uppsala model, where market knowledge is unique to each market. However, even though Johanson and Vahlne (1977) find market specific experience most important for learning about foreign markets, they also acknowledge that the previous experience of other foreign markets is useful. This is also elaborated in the later research of Johanson and Vahlne (1990), where they discuss opportunities of knowledge translation. They suggest that knowledge can be translated, for example, "when the firm has considerable experience from markets with similar conditions, it may be possible to generalise this experience to the specific market." (Johanson and Vahlne, 1990.) This suggests that certain kinds of experiential knowledge can be translated, but with limitations, like "considerable experience", "from markets with similar conditions", etc.

To conclude, previous research suggests that knowledge translation is possible. There are diverging opinions as to how translatable knowledge is, but this can be better understood when knowledge is seen as having an explicit and a tacit component.

Translation of explicit and tacit knowledge

Translation of knowledge is complicated by the fact that it is difficult to distinguish between tacit and explicit knowledge. In the real business world, tacit and explicit knowledge tend to blend in with each other instead of being pure categories. Madhok (1997) describes know-how in a firm as being on a scale from embedded in the capabilities in the firm to generic from one firm to another. Madhok argues that knowledge should be considered a continuous concept, not as a dichotomy. A useful analogy is to see the concept of knowledge as an iceberg (Nonaka, 1994). We can see only explicit knowledge, but tacit knowledge is a very important part. There is both tacit and explicit knowledge in business tasks, and it is primarily the tacit part that influences the possibility of translating (Kogut and Zander, 1993).

Effective knowledge translation depends on the firm's capability embedded in procedures and routines (Makino and Delios, 1997; Lyles and Salk, 1997; Madhok, 1997; Barkema *et al.*, 1996). That is, translation of knowledge determines the performance of firms. An important capability is that knowledge should be useful to the firm. There may be specific knowledge that is not useful outside the realms of the local market. However, there may be knowledge that is useful in several markets, and a firm could benefit from translating such knowledge between markets. This was further studied by Eriksson, Johanson, Majkgård, and Sharma (1997), who found that firms develop capabilities about how to expand internationally, and capabilities associated with specific market relationships. Their findings show that there is an element of internationalisation knowledge that can be applied to any market, and that this knowledge concerns a general knowledge of how to go international. But there is also an element of knowledge about specific markets, which is highly contextual to the local market. Translation of knowledge is thus always difficult because it involves learning about internationalisation in general, and learning about specific markets.

The embeddedness of knowledge being translated

Learning can be analysed by focusing on how past experiences are used in the ongoing business activities. Cohen and Levinthal (1990) argue that the ability of a firm to recognise the value of new, external information, assimilate it, and apply it to commercial ends is critical to its innovative capabilities. They label this capability a firm's absorptive capacity and suggest that it is largely a function of the firm's level of prior related knowledge. Prior related knowledge confers an ability to recognise the value of new information, assimilate it, and apply it to commercial ends. These abilities collectively constitute what they call a firm's absorptive capacity.

Accumulating absorptive capacity in one period will permit its more efficient accumulation in the next. By having already developed absorptive capacity in a particular area, a firm may more readily accumulate what additional knowledge it needs in the subsequent periods. Absorptive capacity is thus critical to exploit any external knowledge that may become available.

In business relationships, the past experiences vary in usefulness in the ongoing business activities. The more useful they are, the more of the firm has developed a capability to relate their prior experiences to ongoing business activities. This capability is a firm's learning ability. The drivers of this process are the current activities taking place in ongoing business activities in order to realise perceived business opportunities.

Conclusion: components of the model of knowledge translation

This chapter has identified three core distinctions for knowledge translation. First is the distinction between tacit and explicit knowledge, second is that between general and market specific knowledge, and the third core is past experiences and present activities. They are related to one another so that tacit and explicit refers to the translatability of knowledge, whereas general and market specific knowledge refers to the embeddedness of knowledge in the local market. Firms can develop their knowledge translation capabilities by relating past experiences to ongoing activities. The model is developed more fully below.

Empirical framework of knowledge translation in international business relationships

Knowledge and learning in international business relationships

It is not easy to translate knowledge in business relationships, but it is a central factor for the development of international business relationships. The idiosyncrasy of a business relationship suggests that much knowledge within a business relationship is tacit. The more parties in a relationship develop their adaptations to one another, the more does their knowledge of how to do business in the relationship become specific to it. The relationship specific knowledge may be more or less prevalent in the multifaceted ties that make up a relationship. The concept of knowledge is thus central for determining what is translated in a business relationship. The distinction between tacit and explicit knowledge provides us with a means to distinguish between what is non-translatable, or specific to a relationship, and what is translatable, or not specific to a relationship. However, knowledge should be recognised as one whole, incorporating both explicit and tacit characteristics. Even explicit knowledge contains its own situational context, which is tacit.

The translation of knowledge in a business relationship is governed by the character of the learning process. In line with Cohen and Levinthal (1990), Hallén, Johanson and Seyed-Mohammed (1991) identify incremental adaptations that are based on what is learned from prior adaptations as the learning process of business relationships. Prior related knowledge is thus essential for learning in international business relationships. Tacit knowledge is mostly specific relationship and is thus important for the firms learning within the business relationship. Explicit knowledge, on the other hand, can be developed in one business relationship and used in the learning processes of other relationships. It can thus be expected that tacit and explicit knowledge have different roles in the learning process of a business relationship.

A frequently discussed issue is that knowledge translation differs within and between firms, and between wholly owned subsidiaries and joint ventures (Kogut and Zander, 1993). However in this framework it doesn't matter, because this paper is concerned with the general knowledge translation model between two different units or organisations, and the focus of this framework is to examine the factors that affect this knowledge translation process.

Knowledge translation model

Knowledge translation takes place within the frame of business relationships. These relationships are most often relationships where the parties have committed much to each other, so that it is a mutual relationship. The parties in such a relationship realise that the joint relationship evolution governs their unilateral development. Needless to say, a mutual business relationship is full of reciprocal interdependencies, and translation of knowledge goes both ways. However, for analytical purposes, a model is formulated where a piece of knowledge is translated from one party to the other in a mutual business relationship.

The model attempts to empirically operationalise a process of knowledge translation. We have previously argued that knowledge cannot simply be divided into tacit and explicit (Madhok, 1997; Nonaka *et al.*, 1994). The explicit is only a small portion of knowledge that can be seen, touched and articulated. However, every piece of explicit knowledge has its own specific situation. That is, when a firm uses knowledge to act in a situation, the new knowledge resulting from that experience contains many situational factors. The knowledge that is tied in with the unique situation of each firm is often more tacit than explicit, whereas the explicit knowledge is often something that can be applied in the unique situation of the firm. Figure 1 shows the division between the explicit and the tacit in a model of knowledge translation. The tacit component is usually larger than the explicit, which is why it is given more space in Figure 5.1.

Soon-Gwon Choi and Kent Eriksson

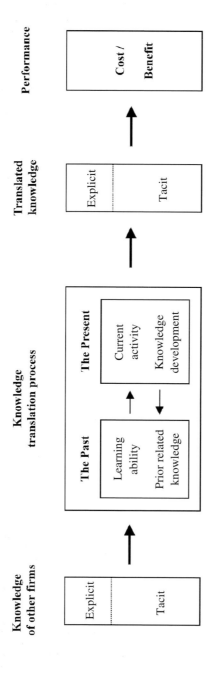

Figure 5.1
Knowledge Translation Process Model

The model distinguishes between the original knowledge being translated and the translated knowledge. Knowledge is usually different in one context compared to another, which means that the translated knowledge is not the same as the original knowledge. The model focuses on knowledge translation, and not what kind of knowledge is being translated, or how the translation process changes knowledge. This is because knowledge has its own situational characteristic. That is, the original knowledge is embedded in the organisation where it resided before translation, whereas translated knowledge will be embedded in the firm that previously did not have that knowledge.

The knowledge translation process consists of a continuing dialogue between past and present (Cohen and Levinthal, 1990). This process can be understood as circular, where past translated knowledge is prior knowledge of the translation process and the translated knowledge will be a prior knowledge of the next translation process. In more detail, the learning ability generated in the firm in the past determines its current activities in the present. The activities done by a firm most often lead to experiences that translate knowledge into that firm. The translated knowledge can then be added to the prior related knowledge, and thus increase the firm's learning ability. The firm can thus acquire knowledge sequentially. The process just described has the implication that organisations cannot acquire all knowledge which they need at once. They can only acquire knowledge which they can understand and absorb. Such understanding and absorption is usually incremental, but sometimes there are more substantial reorientations.

An example of knowledge translation: TQM

A model of knowledge translation is bound to be rather abstract, since there are so many different kinds of knowledge being translated. As a further explanation of knowledge translation processes, we focus on one example: Total Quality Management (TQM) production system (Deming 1986, Ishikawa 1985). TQM is a system of continuously improving quality of production, which is based on the following four assumptions:

"The first assumption is about quality, which is assumed to be less costly to an organisation than is poor workmanship. A fundamental premise of TQM is that the costs of poor quality (such as inspection, rework, lost customers, and so on) are far greater than the costs of developing processes that produce high quality products and services. Although the organisational purposes espoused by the TQM authorities do not explicitly address traditional economic and accounting criteria of organisational effectiveness, their view is that organisations that produce quality goods will eventually do better even on traditional measures such as profitability than will organisations that attempt to keep costs low by compromising quality (Juran, 1974:5.1–5.15; Ishikawa, 1985:104–105; Deming, 1986:11–12). The strong version of this assumption, implicit in Juran and Ishikawa but explicit and prominent in Deming's writing, is that producing quality products and services is not merely less costly but, in fact, is absolutely essential to long-term organisational survival (Deming, 1993:xi–xii).

The second assumption is about people. Employees naturally care about the quality of work they do and will take initiatives to improve it – so long as they are provided with the tools and training that are needed for quality improvement, and management pays attention to their ideas. As stated by Juran (1974: 4.54), "The human being exhibits a distinctive drive for precision, beauty and perfection. When unrestrained by economics, this drive has created the art treasures of the ages." Deming and Ishikawa add that an organisation must remove all organisational systems that create fear – such as punishment for poor performance, appraisal systems that involve the comparative evaluation of employees, and merit pay (Ishikawa, 1985: 26; Deming, 1986:101–109).

The third assumption is that organisations are systems of highly inter-dependent parts, and the central problems they face invariably cross traditional functional lines. To produce high-quality products efficiently, for example, product designers must address manufacturing challenges and trade-offs as part of the design process. Deming and Juran are insistent that cross-functional problems must be addressed collectively by representatives of all relevant functions (Juran, 1969:80–85; Deming, 1993:50–93). Ishikawa, by contrast, is much less system-oriented: He states that cross-functional teams should not sell overall directions; rather, each line division should set its own goals using local objective-setting procedures (Ishikawa, 1985: 116–117).

The final assumption concerns senior management. Quality is viewed as ultimately and inescapably the responsibility of top management. Because senior managers create the organisational systems that deter-mine how products and services are designed and produced, the quality improvement process must begin with management's own commitment to total quality. Employees' work effectiveness is viewed as a direct func-tion of the quality of the systems that managers create (Juran, 1974:21.1–21.4; Ishikawa, 1985:122–128; Deming, 1986:248–249)." Hackman and Wageman 1995: 310–311).

We assume that TQM is translated from Japan to Sweden (Furusten 1995). This is depicted in Figure 5.2, and each step in the knowledge translation model is elaborated below. The original knowledge is the TQM of Japa-nese firms, and the knowledge translation process results in some Swedish version of the TQM, which has a bearing on the cost and benefit of the

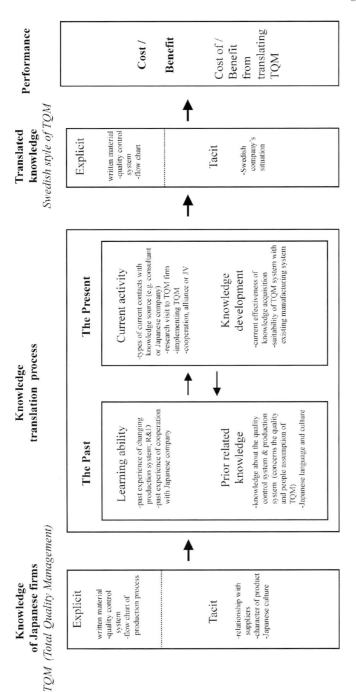

FIGURE 5.2
KNOWLEDGE TRANSLATION PROCESS MODEL (EXXAMPLE)

firm. To a certain extent, the TQM system is made explicit, but there is a considerable tacit uniqueness to both the Japanese TQM and the translated Swedish version.

Past aspects

The past aspects are what the firm has learnt from its experiences. This may concern successful and unsuccessful ventures, where the important aspect is that the firm has drawn some conclusions, or has reflected upon the consequences of their behaviour. The past investment and activities of the firm can be discussed in terms of the firm's learning ability, meaning the ability to accumulate experience from ongoing action, which is determined by how the previously developed knowledge relates to the knowledge being translated.

Prior related knowledge

In all knowledge translation, the starting point to recognise some useful knowledge to translate from other firms, which may simply be when firms see business opportunities. This task hinges upon how knowledge gained from prior experiences relates to the translated knowledge. It is easier to detect a business problem when the knowledge of what to look for already has been developed. In a way, it is difficult to see what is not already known. Prior related knowledge can make it easier to recognise the value of new information and assimilate it. At the most elementary level, this prior knowledge includes basic skills or even a shared language, but may also include knowledge of the most recent scientific or technological developments in a given field (Cohen and Levinthal, 1990).

The role of prior related knowledge in the translation process of TQM systems can be discussed in terms of its four assumptions. The quality assumption stresses the need for a total quality comprehension. Such an insight would no doubt be easier to gain if previous knowledge in a firm is geared towards total quality concepts, instead of, for instance, short-term accounting profit goals. The people assumption states that people are inherently industrious and prone to produce high-quality goods and services. If the firm is built on control and remuneration systems, it will be difficult for them to understand the usefulness of low-control and independent strategies in TQM. A highly centralised and differentiated firm will find it difficult to apply the TQM assumption that teams should be cross-functional and have localised goals. In essence, the TQM system emanates from the Japanese culture and production system. The less prior knowledge a firm has of their culture and way of working, the less effectively will they assimilate the TQM knowledge.

The implications are that the relatedness of prior knowledge has a great effect on the experiences made by firms. Prior related knowledge can also influence the communication costs. Buckley and Casson (1976) insisted that communication costs are crucial in market internalisation process and these costs can vary with the economic, social and linguistic dissimilarities between regions. They also find that it is important to have similar background in encoding and decoding of information in international business. This shows that there are various ways in which prior knowledge can relate to new knowledge.

Learning ability

The firm's learning ability concerns its ability to assimilate the translated knowledge into its structure and routines. The learning ability results from a prolonged process of investment and knowledge accumulation within the firm, and its development is path-dependent. That is, a firm's current learning ability is influenced by its historic participation in specific product markets, lines of R&D, and other technical activities. Accumulating learning ability in one period will permit its more efficient accumulation in the next. Thus by having already developed some learning ability in a particular area, a firm may more readily accumulate what additional knowledge it needs in the subsequent periods in order to exploit any critical external knowledge that may become available (Cohen and Levinthal, 1990). The learning ability of a firm is thus not only the ability to assimilate new knowledge, but also a capability to act in the business environment.

Learning ability differs from prior related knowledge in that it has for instance a more active and flexible characteristic than prior knowledge. Learning ability results from past learning experiences, existing routines and organisational culture. Learning ability can be said to influence the quantity, quality and effectiveness of learning, whereas prior related knowledge only concerns the effectiveness. For example, the past experience of TQM enhances the firm's capability to deal with all sorts of issues that are similar. For instance, the basic assumptions of TQM involve a holistic view on production, market and management quality. Naturally, the learning ability of a firm in these respects will facilitate smooth assimilation of TQM practices, or, conversely, when the firm has assimilated TQM practices they will increase their holistic quality capabilities for future translation situations. The same goes for the other three pillars of TQM: enabling independent staff action; cross-functional involvement; and senior management skills.

Together, the learning ability and the relatedness of prior knowledge make up the past aspects of the knowledge translation model. One simple way to distinguish between them is to consider the learning ability as the ability of the firm, whereas the prior related knowledge is the experience of the firm. Learning ability thus refers to the internal constitution of the firm, whereas prior related knowledge focuses on the external situations which have shaped the firm. However, this process is circular and sequential, even though we discuss the translation of TQM practice as an example. An important implication is that organisations cannot acquire all the knowledge which they need at once. They acquire knowledge most effectively incrementally.

Present aspects

The present aspects are concerned with what and how the firm learns in the present. The fundamental assumption of our model is that the firm's current activities drives the knowledge translation process. It is only by performing activities, either in response to some perceived situation, or pro-actively to capitalize on business opportunities, that the firm relates to past experiences.

Current activity

Current activity is a driver of the knowledge translation. Through current activities a firm can find useful information. That is, firms have to perform activities in order to perceive business opportunities and start to translate knowledge. Current activities are a matter of involvement. As mentioned above, firms have to involve themselves enough to get information, thereby overcoming the limits of their boundaries, making them permeable. This permeability provides organisations with a "window on their partners' broad capabilities" (Hamel, Doz, and Prahalad, 1989; recited from Inkpen and Beamish, 1997). Current activities may be larger events, or repeated smaller interactions in daily business transactions.

Current activity determines the pace and extent of knowledge translation in the TQM case. In order to learn about the TQM system, a firm has to do activities on its own, with consultants, or in cooperation with a firm that has implemented TQM. By trying to implement the four fundamental practices of TQM, the firm finds itself in situations where the results of activities can be reflected upon, that is, where knowledge can be developed. But there is a need for a firm to act in order to be able to develop knowledge. Primarily because knowledge is mostly tacit, it is difficult to develop knowledge in the form of thought experiments. There are too many implicit parameters that are unaccounted for when projections are made in business. Therefore, the holistic assumption of TQM that management, production, and markets should be considered as one is difficult to get to know without trying to act it. Likewise, the knowledge of enabling of independent staff action, cross-departmental involvement, and improvement of senior management skills is difficult to obtain without trying to do it.

Knowledge development

Knowledge development is the process of accumulating knowledge in a firm. It is a process where knowledge is assimilated and applied to commercial ends. Through this step firms can store the new knowledge into their capabilities. Nonaka (1994) emphasises that knowledge development is a complex organisational process involving various levels and actors. Huber (1991) explains this process to be composed of the sub-processes: knowledge acquisition, information distribution, information interpretation, and organisational memory. He stressed especially that demonstrability and usability of learning depend on the effectiveness of the organisation's memory.

The fundamental principles of TQM will all result in knowledge being developed in different areas. If activities are made to increase independence of staff action, then that is an area of knowledge development for the firm. However, the knowledge development in the present is determined by the activities, which in turn are governed by past aspects. So, there is no discerning function in what kind of knowledge will be assimilated in knowledge development *per se*. Rather, the prior experiences determine the nature of knowledge development.

The knowledge developed in the firm is directly dependent on the current activities. The activities lead to a change of situation, which is instrumental for the experiences the firm makes. The current activities are the actions performed *by* the firm, whereas the knowledge development is what is stored *in* the firm. These present aspects will become a prior knowledge and learning ability in the next knowledge translation process.

Cost and benefit of knowledge translation

Knowledge translation requires an effort, and therefore incurs costs. The costs can be reduced if the firm's past experiences help the firm to avoid some pitfalls in pursuing perceived business opportunities. Firms that have developed a learning ability and can relate their prior knowledge to the current activities know more about how to translate knowledge. They can thus avoid excessive costs and increase benefits in terms of both money and increased learning abilities.

Conclusion of the knowledge translation process

Knowledge translation is driven by current activities. That is, the commitment to translate knowledge is determined by the current activities. But the past experiences determine the depth of knowledge involved in the activities. The more experienced firms will perform current activities that are more well suited for the market. However, inexperienced firms can develop their capabilities to translate knowledge by committing more resources to current activities.

Conclusions

Knowledge translation is a difficult phenomenon to analyse. It is a very important problem and crucial to organisations in this era of international competition. But it is elusive, since knowledge has many faces and it is therefore often difficult to make anything other than very general models. This paper has developed a model for an empirical research framework for knowledge translation in international business. The model identifies the translation process as the focus of study. It thereby recognises that all business situations are unique, and that the processes going on in these unique situations have common properties.

In particular, knowledge is treated as having both unique and more general properties. Many researchers divide the characteristics of knowledge into two dimensions, tacit and explicit. However, it is better to understand knowledge as a whole. The firm is always in a unique business context, and the knowledge it develops is inseparable from this context. Knowledge contains both tacit and explicit dimensions at the same time.

The model also identifies the interplay between the firm's past experiences and its current activities as crucial for knowledge translation. Knowledge can transfer; however, it takes time and is costly. These costs vary very widely due to factors in the translation process model. In that model, two things are important: past and present aspects. The cost of knowledge transfer doesn't depend only on the current situations. Past situations influence also decisions about costs and time. Thus it is important to consider both sides in researching knowledge transfer.

An interesting reflection on the interplay between the past and the present is to what extent the knowledge development process is deterministic. If the past determines not only the future activities, but also the experiences a firm has, then the path dependence would be strong and deterministic. However, our knowledge translation model emphasises that current activities lead the firm to develop new knowledge. To a certain extent, this new knowledge is contingent on what the firm already knows, but the firm also recreates much of its reality in the face of activities it performs. Past events are continuously reflected upon and re-interpreted by firm staff, and it is as representations in present activities that these past experiences are carried forth. Therefore, even though there is an element of path dependence in knowledge development, the strength of it may be debated.

However, path dependence is evident in the incrementality of the knowledge translation process. Knowledge translation cannot happen only once. Instead the relatedness of prior knowledge is instrumental to what the firm learns from its current activities. This process is circular and sequential. Continuous communication between past and current aspects is the crucial point of the knowledge translation process.

Managerial implications

Knowledge translation is an essential factor in international business. Two common issues are that the comparative advantages of other firms must be learned and that the firm's own competence must be translated all over the world. However, the translation process does not happen at once, and needs to be given time. The managerial implications of the knowledge development model are that firms need to recognise the properties of the process, and act accordingly.

For instance, knowledge translation is often a path-dependent process because it demands past investment and experience. Faced with business problems, the firm's prior related knowledge and past translation experience will play a crucial role. If organisations do not have related prior knowledge and experience, it will take more time and cost more. Thus every firm has to invest continuously in R&D, and has to learn and try to assimilate others' comparative advantages. In particular, a firm should take care to recognise that the knowledge it has articulated is tied in with more tacit knowledge which is implicit in its way of pursuing its work. If staff understand more about knowledge translation processes, they will increase the firm's sensitivity to its business. Such sensitivity helps to avoid excessive mistakes and reduces costs.

Second, we have to remember that the process of translation is an incremental process. Changes cannot happen in the short term and are a continuous communication between firm capabilities and the source of the translation. Therefore, knowledge translation demands a long-term perspective, and continuous investment and interest. In particular, radical re-orientations will often fail since the firm has a poor sense of what to do since their activities are insensitive to the business situation. Therefore, forced development requires special attention to the learning by staff in the firm. Alternatively, the firm can resort to acquiring existing ventures and try to incorporate them. However, acquiring another firm does not reduce the growth problems, it merely moves them from being market related to being inter-firm related. The knowledge translation process remains the same in either case.

Third, the results of this research point to a need for strategic management of business relationships. Knowledge translation is made more efficient in business relationships with a long history and a deep level of involvement. So managers in firms have to choose the relationships with other organisations and try to build and capitalise on the strength of the contacts. That is, they have to decide the extent of the relationships with other organisations (strategic alliance, licensing, etc.) to match the knowledge translation in the relationship. This also means that firms should be sensitive to changing their formal mode of cooperation to better suit the depth of knowledge translation between them.

Finally, the knowledge translation process requires proper leadership and organisational culture. A preferred situation is where the leadership and organisational culture encourage learning from diverse markets. In particular, sensitivity to the business processes should prevail in the firm. This sensitivity can influence the level of cost and benefit of the translation process.

References

ANDERSEN, OTTO 1993, On the internationalization process of firms: A critical analysis *Journal of International Business Studies*, 24(2): pp. 209–32.

ANDERSON, E. and WEITZ, B. 1992, *The Use of Pledges to Build and Sustain Commitment in Distribution*.

ALTER, C. and HAGE, J., 1993, *Organizations Working Together*, Newbury Park, CA: SAGE.

AXELROD, R. 1984, *The Evolution of Cooperation*, New York: Basic Books.

BANDURA, A., 1977, *Social learning theory*, Englewood Cliffs NJ: Prentice-Hall.

BARKEMA, H. G., BELL, JOHN H. G, and PENNINGS, JOHANNES M., 1996, Foreign entry, cultural barriers, and learning *Strategic Management Journal*, Vol. 17, pp. 151–166.

BILKEY, WARREN J. and TESAR, GEORGE 1977, The export behaviour of smaller sized Wisconsin manufacturing firms *Journal of International Business Studies*, 8(1): pp. 93–98.

BLOIS, K. J. 1972, Vertical Quasi-Integration *Journal of Industrial Economics*, Vol. 20(3), July, pp. 253–272.

BUCKLEY, PETER, and CASSON, MARK, 1976, *The future of multinational enterprise*, London: Macmillan and Co.

CARLSON, SUNE, 1974, International transmission of information and the business firm *The Annals of the American Academy of Political and Social Science*, 412: pp. 55–63.

CAVUSGIL, S. TAMER, 1984, Organizational Characteristics associated with export activity *Journal of Management Studies*, 21(1): pp. 3–22.

COHEN, WESLEY M. and LEVINTHAL, DANIEL A., 1990, Absorptive Capacity: A new perspective on learning and innovation *Administrative Science Quarterly*, 35, pp. 128–152.

CZINKOTA, MICHAEL, 1982, *Export development strategies: US promotion policies*. New York: Praeger Publishers.

DEMING, W. EDWARDS, 1986, *Out of the Crisis*, Cambridge, MA: MIT Center for Advanced Engineering Study.

DEMING, W. EDWARDS, 1993, *The New Economics for Industry, Government*, Cambridge, MA: MIT Center for Advanced Engineering Study.

DODGSON, MARK, 1993, Organizational Learning: A Review of Some Literature, *Organization Studies*, 14/3, pp. 375–394.

ERIKSSON, K., JOHANSON, J., MAJKGÅRD, A., and SHARMA, DEO, 1997. Experiential Knowledge and Cost in the Internationalization Process *Journal of International Business Studies*, 28(2).

FORD, D., (ed.) 1990, *Understanding Business Markets: Interaction, Relationships and Networks*, San Diego: Academic Press.

FORSGREN, MATS, 1997, The Advantage Paradox of the Multinational Corporation in *The nature of the international firm; Nordic contributions to international business research*, eds Ingmar Björkman and Mats Forsgren, Copenhagen: Handelshojskolens Forlag.

FURUSTEN, S., 1995, The Managerial Discourse – A Study of the Creation and Diffusion of Popular Management Knowledge, diss., Department of Business Studies, Uppsala University.

HACKMAN, J. RICHARD and WAGEMAN, RUTH, 1995, Total quality management: empirical, conceptual, and practical issues *Administrative Science Quarterly*, Jun 95, Vol. 40, Issue 2, p. 309, 35p.

HALLÉN, L., JOHANSON J., and SEYED-MOHAMED, N., 1991, Interfirm adaptation in business relationships *Journal of Marketing*, 55(April), pp. 29–37.

HAMEL, G., DOZ, Y. L., and PRAHALAD, C. K., 1989, Collaborate with your competitors – and win *Harvard Business Review*, 67(1): pp. 133–139.

HAMEL, G. 1991, Competition for Competence and Interpartner Learning Within International Strategic Alliances *Strategic Management Journal*, Vol. 12, pp. 83–103.

HEDLUND, GUNNAR, 1994, A Model of Knowledge Management and the N-form corporation *Strategic Management Journal*, Vol. 15, pp. 73–90.

HIPPEL, E. VON, 1988, *The sources of innovation*, Cambridge: Cambridge University Press.

HOBDAY, M., 1990, *Telecommunications in developing countries: the challenge from Brazil*, London: Routledge.

HUBER, GEORGE P., 1991, Organizational learning: the contributing processes and the literature *Organization Science*, 2(1), pp. 88–115.

INKPEN, ANDREW C., and BEAMISH, PAUL W., 1997, Knowledge, bargaining power, and the instability of international Joint Ventures *Academy of Management Review*, Vol. 22, No. 1, pp. 177–202.

ISHIKAWA, KAORU, 1985, *What is Total Quality Control? The Japanese Way*, Englewood Cliffs, NJ: Prentice-Hall.

JOHANSON, J., and VAHLNE, J.-E., 1977, The Internationalisation Process of the Firm – A Model of Knowledge Development and Increasing Foreign Market Commitments, *Journal of International Business Studies*, 8, No. 1, Spring/Summer, pp. 23–32.

JOHANSON, J., and VAHLNE, J.-E., 1990, The Mechanism of Internationalisation, *International Marketing Review*, 7, No. 4, pp. 11–24.

JURAN, JOSEPH. M., 1969, *Managerial Breakthrough: A new concept of the Manager's Job*, New York: McGraw-Hill.

JURAN, JOSEPH. M., 1974, *The Quality Control Handbook*, 3rd ed., New York: McGraw-Hill.

KOGUT, BRUCE, and ZANDER, UDO, 1993, Knowledge of the firm and the evolutionary theory of the multinational corporation *Journal of international business studies*, fourth quarter, pp. 625–645.

LATOUR, BRUNO, 1986, The powers of association in *Power, Action and Belief, A New Sociology of Knowledge?* ed. John Law, London, Boston and Henley: Routledge & Kegan Paul.

LYLES, MARJORIE A., and SALK, JANE E., 1997, Knowledge acquisition from foreign parents in international joint ventures; An empirical examination in the Hungarian context, *In Co-operative strategies; European perspectives*, eds Paul W. Beamish and J. Peter Killing, San Francisco: The New Lexington Press.

MADHOK, ANOOP, 1997, Cost, value and foreign market entry mode: The transaction and the firm *Strategic Management Journal*, Vol. 18, pp. 39–61.

MAKINO, SHIGE, and DELIOS, ANDREW, 1997, Local knowledge transfer and performance: Implications for alliance formation in Asia in *Co-operative strategies; European perspectives*, eds. Paul W. Beamish and J. Peter Killing, San Francisco: The New Lexington Press.

MODY, A., 1990, *Learning through alliances*, Washington: The World Bank.

MOWERY, DAVID C., OXLEY, JOANNE E., SILVERMAN, BRIAN S., 1996, Strategic alliances and interfirm knowledge transfer *Strategic Management Journal*, Vol. 17 (Winter Special Issue), pp. 77–91.

NONAKA, IKUJIRO, 1994, A dynamic theory of organisational knowledge creation *Organization Science*, Vol. 5, No. 1, February, pp. 14–37.

NONAKA, IKUJIRO, BYOSIERE, PHILIPPE, BORUCKI, CHESTER C., and KONNO, NOBORU, 1994, Organizational Knowledge Creation Theory: A first comprehensive test, *International Business Review*, Vol. 3, No. 4, pp. 337–351.

O'GRADY, SHAWNA, and LANE, HENRY W., 1996, The psychic distance paradox *Journal of International Business Studies*, 27(2): pp. 309–333.

POLANYI, M., 1966, *The tacit dimension*, London: Routledge & Kegan Paul.

REID, STAN. 1983, Firm internationalization, transaction costs and strategic choice. *International Marketing Review*, 1(2): pp. 44–56.

ROSENBERG, N., 1982, *Inside the black box: technology and economics*, Cambridge: Cambridge University Press.

SPENDER, J.-C., 1996, Making knowledge the basis of a dynamic theory of the firm, *Strategic Management Journal*, Vol. 17(Winter Special Issue), pp. 45–62.

STINCHCOMBE, A., 1990, *Information and organizations*, Berkeley: University of California Press.

TOYNE, B., 1989, International exchange: A foundation for theory building in international business. *Journal of International Business Studies*, 20 (Spring), pp. 1–17. Channels', *Journal of Marketing Research*. Vol. XXIX, pp. 18–34.

Part II Network Relationship Learning

CHAPTER 6

The Transferability of Knowledge in Business Network Relationships

KENT ERIKSSON and JUKKA HOHENTHAL[1]

Introduction

Markets may be considered to be networks of business relationships (Nohria and Eccles 1992, Anderson, Håkanson and Johanson 1994, Powell, Koput and Smith-Doerr 1996, Porac, Thomas and Baden-Fuller 1989), as a consequence of which, the business relationship is the pivotal unit of analysis. Much research has focused on developing models for the study of business relationships (Levinthal and Fichman 1988, Morgan and Hunt 1994, Alter and Hage 1993, Mayer, Davis, and Schoorman 1995, Anderson and Weitz 1992, Ring and Van de Ven 1992; 1994, Håkansson and Snehota 1995) and case studies of business relationships in business networks often reveal a complicated pattern of interrelated factors (Waluszewski 1989, Sjöberg 1996). Despite the considerable amount of research that has been undertaken, it would appear that there are no such models that discuss the transferability of knowledge within business network relationships. This study is intended to remedy this gap in our understanding of business network relationships by presenting a framework for the analysis of the idiosyncrasies of business relationships.

A key for understanding the development and transferability of knowledge within and between relationships is to understand the nature of the technology involved in the business relationships. Technology has often

---

[1]The authors appear in alphabetical order, and have made equal contributions to this paper.

been seen as something that is exogenous to individual firms, as a paradigm of production possibilities to be used for efficient production. Contingency theory considers technology as something that determines the structure of production (Woodward 1965, Lawrence and Lorsch 1967). However, we argue that this claim overlooks the fact that the technology and the structure evolve reciprocally over time (Barley 1986, 1990). We define technology as the set of tasks by which input is acted on in order to produce output (Engwall 1978). Technology can be considered as a way to organise the work flow. And this organising can be performed as part of a conscious plan, but it can also be an emergent structure that is developed as a result of action and interaction.

Technology adaptations are often made in response to relationships between firms as the customer and seller have to adapt their products to each other to gain efficiency. This is a more or less automatic process as the parties learn to take advantage of each other's knowledge and thus gradually change their organisation of the interaction to create a better match between them (Hallén, Johanson, Seyed-Mohamed 1991). Since each business partner has its own unique business setting, relationship building incorporates a considerable degree of relationship specific, or idiosyncratic, investment that has little or no value outside the context of the relationship. Anderson and Weitz (1992) found idiosyncratic investments to be the primary drivers of relationship building, however, there is little understanding of the nature of these idiosyncratic investments. This suggests that a closer examination of the process by which technology leads to the adaptation of relationships could be a fruitful way to improve the understanding of the nature of idiosyncratic relationship investments. In presenting an initial observation that technology and adaptations are associated, the study by Hallén et al. (1994) pointed to the need for the development of a framework for the study of technology in relationship adaptation.

These idiosyncratic relationship investments are, however, not done in isolation. All changes in individual relationships are made with the total network of business relationships as a background. Thus change in one relationship is dependent on and will inevitably have an influence upon the firm's other relationships: The technology developed in one business relationship will influence the technology in other relationships, and changes in technology in one relationship will be dependent on changes in yet other relationships. The understanding of how technology adaptations in one relationship are contingent on adaptations in another is a key to understanding the dynamics of knowledge transfer between business relationships in a network. This paper intends to fill a gap in research by studying idiosyncratic relationship specific knowledge and knowledge transferable to other relationships.

The purpose of this paper is to develop a framework for understanding the relation between idiosyncratic relationship specific knowledge and knowledge transferable to other relationships.

In order to fulfil this purpose, we propose that technology can be seen to incorporate different degrees of tacitness where a high degree of tacitness corresponds to a specific situation, and a low degree can be applied to many situations. The degree of tacitness is the tool with which the nature of relationship adaptations will be analysed.

Adaptations in business relationships

Business life revolves around the buying and selling of products and services. Empirical studies have revealed that buyers and sellers do business in relationships that last for long periods of time and which they take considerable pain to develop (Ford 1990). Thus individuals working in these relationships spend a large amount of their time and working on the exchange of goods and services in the relationship. Consequently, what is learnt by the individuals involved in buying and selling is associated with their exchange activities and has substantial implications for the rest of the organisation. While any organisation may be involved in many activities, those that are significant for its business exchange are probably focused on the exchange of goods and services.

It has been observed that the exchanges going on in a relationship are not discrete transactions, but rather, they are part of a process of mutually beneficial development of the relationship (Hallen, Johanson and Seyed Mohamed 1989). Over time, and once the parties have committed themselves to the development of the relationship, it has evolved into something much more than just a way of transacting between parties. In the initial stages the parties make pledges in order to feel each other out, and the result of the pledge making is most often a commitment to do business with each other. The pledges are made in the form of investments devoted to the development of the relationship (Anderson and Weitz 1992). As the parties turn the pledge making into a concrete exchange of goods and services, they also start to transform the business exchange into a business relationship.

Anderson and Weitz (1992) found that idiosyncratic investments are more important than contractual terms as drivers of the pledge making process. Idiosyncratic investments are relationship specific and have little or no value outside a specific relationship. An explanation for this is that the investments involve any use of firm specific resources intended to develop the business relationship between partners. In this context, resources are widely defined to comprise tangible and intangible assets, such as the unique capabilities of each firm in the relationship (Collis 1996). As firms invest to do business with each other they create new capabilities within the relationship. These relationship specific capabilities are initially based on the capabilities of the firm; however as the relationship capabilities gradually evolve they become different from the firm's capabilities as a whole.

Once the exchange between parties has started, the firms increase their knowledge about each other, as the exchange activities between them are co-ordinated (Alter and Hage 1994). Such knowledge may include policies related to the holding of stock, product development, production processes, payment routines, etc. As a result the parties gain a deep knowledge of each other that surpasses mere information required to exchange goods and services. The parties learn how to match their abilities much better and teach each other about their respective needs and abilities (Hallén, Johanson and Seyed-Mohammed 1991). Matching within a relationship is obtained through the activities that parties make in relation to each other. The adaptations and modifications of products and routines connected to the exchange will eventually transform the relationship and give it a "quasi-organisational" character. This leads each party to consider the parts of the exchange partner's activities that are connected to their own activities as being as important as the activities going on inside the organisation.

The nature of adaptations can be illustrated by an example provided by Tyre and von Hippel (1997) who describe the adaptations made when a machine that places components on a circuit board was delivered to a customer. Shortly after installation, the user observed that the accuracy with which the components were placed was lower than anticipated. The user described the symptoms to an engineer, but the engineer could not understand the problem and visited the plant where the machine had been installed. Even though the engineer saw the machine in context, he could not locate the problem. After several trips back and forth to the plant, and after using diagnostic tools and after consulting with colleagues and with other buyers, the problem was located. A coupling was worn. The coupling withstood wear in the laboratory tests, but in actual use, it did not. The supplier made an improvement to the coupling and installed it in the machine.

Even though the actual installation of the machine is only one part of the sequence of events that make up a business relationship, it still illustrates the complexities involved in working out issues in the relationship. The installation of the machine highlights the role of the physical context for relationship evolution. Machines are often thought of as being decoupled from the idiosyncrasies of a relationship, but this example illustrates how technology can be highly contextual and idiosyncratic.

Degree of tacitness in technology

The adaptations made in exchange relationships will thus make the technology more contextual and relationship specific. At the same time the relationship technology will be influenced by the relationship specificity of surrounding technologies. Technology can be considered as a means by which to organise the work flow. It involves everything made in order to transform inputs into outputs, both within an organisation and in the quasi-organisation of the relationship. This organisation can be a conscious plan, but it can also be an emergent structure that is developed as a result of interaction.

Hutchins (1991) gives an interesting example of how work is reorganised in response to a change in the informational environment. While entering a harbour, a large ship suffered an engineering breakdown that disabled an important piece of navigational equipment. The ship's navigation team responded to the changed demands imposed by the loss of equipment. Following a rather chaotic search for alternatives – computational and organisational – the team arrived at a new temporarily stable work configuration. This solution was discovered by the organisation itself before it was discovered by any of the participants.

The result of the organising activity performed by the individual actors is a stable collective structure for organising the work flow; a technology for handling the situation. This technology is partly tacit and partly explicit. Tacit knowledge is developed when the actor works toward a specific goal (Polanyi 1969) and thereby creates an understanding for the variation in the problems and the set of actions – the patterns of actions – needed to solve them. Explicit knowledge is the codified part of the action pattern of the individual actor. The division into tacit and explicit knowledge is always individually constructed in the process of doing something. Technologies demanding more tacit knowing demand that individuals whose actions make up the technology have to develop tacit knowing to be able to carry out the activities.

Organising implies the institution of a chain of interlocked activities (Weick 1979) which can contain a larger or smaller tacit component. The larger the tacit component, the more difficult it becomes to exchange an actor involved in that particular technology. All activities contain both a tacit and an explicit part. We can usually say something about what we do, but we cannot say everything. Tightly knit interlocked activity chains containing a large tacit part are an efficient way of organising the work flow. This is especially true when there is need for swift action without time for reflection. One has to act without thinking, automatically implementing difficult and complex patterns.

Other actors can use this effective action pattern to attain certain goals. When using a forwarding company or a law firm, we make use of their action patterns to solve problems concerning distribution or legal matters. It would require a considerable investment that will constrain and empower him, making some action patterns possible while hindering others. and redevelopment of competence to create the system of inter-locked activity patterns that these firms offer. An actor will take it for granted that the surrounding structure can create the inputs needed and give the required output. The actor is thus embedded in a system of inter-locked activity patterns.

We propose that the nature of technology can be better understood if one bears in mind its tacitness and explicitness. Following Zander (1991), degree of tacitness can be divided into the concepts: teachability, articulat-edness, observability, and complexity. It can be argued that almost by definition tacit technologies are less teachable, less well articulated and easy to observe in use, complex and systemic. Tacit technologies are more difficult to learn for the individual, meaning that tacit technologies are more difficult to transfer to other firms (*cf.* Williamson 1975). Thus the transfer of tacit knowledge is a matter of transferring activity patterns from one context to another.

Teachability focuses the amount of supervision needed to teach someone to do a task. It has also to do with access to the interlocked activity chain in the firm. Skills containing a large tacit part are more diffi-cult to teach than skills with a large explicit part. Skills that are tightly interlocked with the performance of other skills are only teachable if the actor has access to these specific skills. An actor who becomes part of a tightly interlocked sequence of actions will have to learn through inter-acting with the other actors.

Articulatedness is dependent on a standardised, controlled context for the performance; on a performance that can be cut down to a set of simple parts that relate to one another in very simple ways (Zander 1991: 117–120). Writing a manual often means that the action sequence is chopped up into easily manageable parts, but the actor still has to put these parts together in a continuous flow to become a knowledgeable actor. Tacit knowledge expressed in the form of easily manageable parts of a larger activity sequence would make it easier to transfer the activity between one context and another. It would also imply a reduction in the tacitness and variation and thus mean that the activity is more stable.

Observability has to do with how easy it is to understand the manufac-turing process by looking at and examining different aspects of the manufacturing process or the final product (Zander 1991: 145–151). Using standardised equipment and following strict rules makes activities more observable in the work flow. Thus sequences of activities with little variation that use straightforward equipment would imply the existence of a low degree of tacit knowledge and sequences of activities that are easy to transfer. It would also imply more rigid activity sequences that are difficult to change.

Complexity has to do with the amount of information required to characterise the knowledge in question (Winter 1987). A high degree of complexity demands extensive education, training and experience with the technology as well as close contacts between individual actors, leading to interlocking behaviour patterns with a high degree of tacitness. This would result in an activity sequence that is relatively easy to change since it already contains plenty of variation.

The more tacit the technology of a particular activity pattern, the more difficult it would be to transfer and the easier it would be to change. A tacit activity pattern can handle more variation at the same time as it is more complex and difficult to transfer to another context.

The technology in adaptations

Thus a key factor in business relationships is the technology used to produce the goods and services. For instance, a study found differences in how firms adapted to each other depending on whether their technology was unit, mass or process production (Hallén, Johanson, and Seyed Mohamed 1994). However, in the study just mentioned technology is considered to be homogenous within a relationship. Several studies have identified that one organisation may contain many technologies (*cf.* Scott 1981), and analogously it can be expected that a business relationship may incorporate several different kinds of technologies. The technology involved in one business relationship will be dependent on the technology required for the different activity patterns the firm is involved in.

A framework developed by Engwall (1978) splits a firm's technology up by identifying that different workflows in a firm can differ in terms of the technology used for operations, raw materials or knowledge involved in production. Operations technology concerns the actions replaced by machines in workflow activities and can be subdivided into three categories: a) automation, i.e., the extent to which human actions are replaced by equipment. b) workflow rigidity, i.e., the interdependence and interchangeability of segments of the operations sequence; and c) specificity of evaluation, i.e., the possibility of evaluating the operations performed. Materials technology concerns the nature of the raw materials acted upon, of which there are two subcategories: a) understandability, how the organisation understands the nature of the raw material; and b) variability, i.e., the extent to which an adjustment must be made to the raw material. Knowledge technology concerns the complexity of the knowledge used in the workflow. It can also be divided into two categories: a) number of exceptional cases; and b) analysability of the problem in terms of the nature of the search process undertaken when special cases occur.

Technology has traditionally been considered to be more or less objective for each task. It governs how work is organised. It can therefore be expected that different dimensions of technology can have different roles when adaptations occur to business relationships. The knowledge involved in business relationships can therefore be classified in terms of degree of tacitness or explicitness by the type of technology as in Table 6.1 below.

TABLE 6.1

A framework for analysing degree of tacitness in technology in business relationship adaptation

	Degree of Tacitness in Technology	
	High	**Low**
Operations Technology Automation Workflow rigidity Specificity of evaluation		
Materials Technology Understandability Variability		
Knowledge Technology Number of exceptional cases Analysability		

The application of the research framework to an empirical context may provide further clues to the nature of the knowledge involved in international business relationships. The earlier discussion of the tacitness of technology suggested some more concrete dimensions that can be used. These concern whether the technology is teachable, articulable and observable in use, and whether it is simple. Each dimension of technology can be analysed with regard to the four dimensions of tacitness. This will give the possibility of reaching a level of analysis that is closer to the operative level, which makes it easier to assess such abstract concepts as tacitness and technology. The results of all aspects of specific dimensions of technology will provide a basis for assessing the knowledge of the entire adaptation process in a business relationship.

Automation in Table 6.1 concerns the actions replaced by machines in the workflow. It may be expected that the knowledge associated with machinery could be found in manuals, which, together with training would help to teach the firm about how the machinery works and what it does. It is also probable that automation makes the technology observable in use, and also that the components involved are reasonably simple and can be defined independently from parts treated by machines. In all, it seems more likely that automated technology has a low degree of tacitness.

It should be emphasised that automation does not solely refer to the machinery used in production. While the introduction of machinery may create new jobs, such as the monitoring of the machine, the technology is automated because machines can be made to perform tasks that would otherwise be done by people. For example, the monitoring of processes in paint production is a highly complex task that requires an understanding of the complex production processes. A monitor may use computerised images to survey several production sequences, which requires him to envision how the sequences are linked and how the machinery works in reality (Weick, 1990).

However, the fact that machines consist of physical parts suggests that the work procedures have been made less tacit. Long before machines were used in paint production, a master may have used his experience when assessing the mixing of ingredients for the paint. Perhaps he rubbed paint powder between his fingers to determine the quality of the input and then fine-tuned each batch of paint produced by adjusting the quantities of some components. By adding machinery, what was previously in the master's head and in the actions he performed becomes quantified in rigid quantities and mechanical movements. Technology is concerned with the workflow and not with some perceptions of the machinery.

Therefore automation involves making technology less tacit for the actors involved. Automation will probably lead to a technology that is more teachable, articulated and observable in use, and less complex. This would imply that a more automated workflow is easier to transfer and more difficult to change. It is also more systemic, making the larger system less tacit.

Workflow rigidity concerns the interdependence and interchangeability of tasks. It can be argued that work tasks that have a fixed function in relation to each other need articulated work procedures. There may be a need for workers to clearly perceive that one task needs to be completed before another. If it is necessary to articulate this, then it will be necessary to teach those operating the workflow. For the same reasons, it can be inferred that rigid workflows create a need to make each task simple and explicitly defined in relation to others. In a rigid workflow there is less room for variation. Returning to the paint production example, if the sequence of activities in the workflow is rigid in relation to each other, then their interrelatedness has probably been articulated in routines, procedures or manuals. Rigid workflows are probably also more observable in use since they involve less variation. Rigid workflows imply that the knowledge involved is an integral part of a larger system, which suggests a high degree of tacitness, but on balance a rigid workflow may involve a lower degree of tacitness. A rigid workflow will create a stable configuration of activity patterns that is easier to transfer and more difficult to change.

Workflow rigidity is similar to the concept of independence that Zander mentions as a means by which to assess the degree of tacitness. Just like workflow rigidity, independence refers to the strength of coupling between activities. The other aspect of this dimension is interdependence; the more interdependent two activities are, the higher the level of tacitness since two sets of more or less tacit activities are interlocked through a more or less tacit process.

A highly specified evaluation of workflow sequences suggests that the source of trouble can easily be identified when something goes wrong. With this in mind it can be expected that there exist procedures for assessment when something goes wrong. These procedures have probably also been taught to the staff who have to do the evaluation. A specified evaluation means that it is possible to make specifications of required inputs and outputs. This is more likely to happen when the knowledge involved is simple rather than complex. By similar reasoning, the possibility of specifying the evaluation of a workflow sequence suggests that some measure of independence can be observed vis-à-vis other workflow sequences. Similarly, high observability of evaluation may imply a lower degree of tacitness. In the example, paint production may have a low degree of tacitness since the colour of the paint is observable and can be compared to standard colours. The reasoning points to specified evaluations having a low degree of tacitness.

The understandability of raw materials used in production in a relationship can also be linked to knowledge. Easily understandable raw material probably involves knowledge that is simple, articulable and teachable. The character and function of understandable raw material may be described in everyday terms and/or using some specific terms of the trade. It is also likely that understandable raw material is observable in use and independent of the production process since it can be observed as one of several inputs. Together, this makes it likely that an understandable raw material will have a low degree of tacitness. Using the paint production example, the production probably involves mixes of specified qualities of chemicals, metals and other kinds of powder. It is likely that each of these components has complicated relations to other components, but this complexity is contained within an automated, rigid technology.

The variability refers to the degree to which adjustments need to be made to the raw material. If the raw material varies very little, then it may be simple rather than complex. Little variation may also make it possible to distinguish the characteristics of the raw material, thus making it articulable and teachable. Materials that vary little can be counted upon to have the same effect in the input to workflows, which can make it possible to treat them independently from the workflow. Because they do not introduce variations to the system, they may also be easier to analyse. The observability of a raw material input that varies a great deal may be much harder than the observability of one that does not. All in all there seems to be support for the argument that a small variation in raw materials necessitates knowledge with a low degree of tacitness. For instance, paint production based on raw material of a consistent quality, such as synthetic chemicals, can probably be articulated in manuals, as it is easy to analyse and more observable in use than paint production based on raw materials that vary considerably, such as a fungus found in nature. Easily understandable raw materials with little variation imply a low degree of tacitness, making this part of technology easy to transfer, but difficult to change.

The first dimension of knowledge technology is the number of exceptional cases involved in the workflow. If there are few exceptional cases that do not fall within routines, then it is likely that the knowledge involved is simple rather than complex. Few exceptional cases may also mean that the knowledge involved in production is independent of other parts of the system since little disturbance occurs. Few exceptional cases mean that there is some stability in the knowledge involved. This can make it easier to articulate and teach it, but it does not necessarily do so. Production that is observable in use usually involves less exceptional cases as it can be monitored and analysed. On balance it seems likely that few exceptions are associated with a low degree of tacitness of knowledge. Paint production that involves few exceptional cases is probably also reasonably articulable and observable.

Easily analysable work tasks are likely to be reasonably simple and independent from complicating interdependencies with other tasks. Being easily analysable, they are probably also articulable and teachable. The observability of use is not readily applicable in relation to the analysability of work tasks. There seem to be reasons to consider analysable work tasks as including a low degree of tacitness. Technology involving less exceptional cases that is easy to analyse has a low degree of tacitness, making it easier to transfer, but more difficult to change.

TABLE 6.2

Working hypothesis for knowledge involved in technology for adaptations in business relationships

	Degree of Tacitness in Technology	
	High	**Low**
Operations Technology Automation Workflow rigidity Specificity of evaluation	Low degree of automation Flexible workflows Unspecified evaluations	High degree of automation Rigid workflows Specified evaluations
Materials Technology Understandability Variability	Incomprehensible Much variation	Easily understandable Little variation
Knowledge Technology Number of exceptional cases Analysability	Many exceptions Not analysable	Few exceptions Analysable

Kent Eriksson and Jukka Hohenthal

The detailed discussion of degree of tacitness in relationship adaptations provides the grounds for formulation of the hypothesis that certain kinds of technologies have different degrees of tacitness. More explicit knowledge is probably often associated with an operations technology that is automated, and has a rigid workflow and specified evaluation techniques. The materials are likely to be easy to understand and to vary little. Knowledge technology is characterised by few special cases that are easily analysable when they occur. Even though cases will be found that diverge from this, it still provides an argument that business relationships with this kind of technology are more likely to have a higher degree of explicit than tacit knowledge. The knowledge involved may thus not be as relationship specific as in relationships where adaptations involve more of tacit technologies.

Adaptations of the technology

As mentioned earlier, several observations support the view that technology is not an objective determinant of adaptation. Instead, the effect of technology is highly contextual. In relationship adaptations, the context of the technology made up of the two firms and the relationship. The learning ability of the firms involved in the relationship is important since the firm that learns more easily has a better chance to incorporate tacit technologies into its internal routines and thereby facilitate adaptations in the relationship.

What we have been discussing is a form of adaptive learning where people act, then interpret and evaluate the response from the environment (March 1994). The environment consists of other individuals and their responses to our actions. Locally instigated change may therefore have unanticipated consequences for other parts of the system that were not part of the design. Locally designed change provokes local adaptations by other parts of the system. These local adaptations are partly tacit and partly explicit.

Tacit knowledge is difficult and costly to transfer from one actor to the other since it has to be translated by the receiving actor to fit his/her knowledge. Individuals develop their knowledge as they act, so learning to do something does not mean that we can describe how to do it in some way other than by demonstration (Polanyi 1969). Tacit knowledge is even more problematic at an organisational level. Organisations consist of multiple actors with interlocking action sequences triggered by each other (Weick and Roberts 1993). Each actor is a carrier of a part of the tacit knowledge needed to carry out the collective activity. Organisations store individual tacit knowledge in the form of an organisational memory consisting of technological and cognitive tacit knowledge (Levitt and March 1988, Nonaka and Takeuchi 1995, Walsh and Ungson 1991). The degree of tacitness is connected to the individual working with a technology, but the degree of tacitness in the technology is a property of the firm and the relationship.

The effect of tacitness of technology on business networks

When two actors interact within the framework of a business relationship, contingent on the networks within which these actors are embedded, they develop common knowledge that can be used by both parties. This knowledge is partly relationship specific and partly transferable to other relationships. Knowledge with a low degree of tacitness, a high degree of automation, rigid workflows, specified evaluations, easily understandable materials technology with little variation, few exceptional cases and analysable technology is easy to transfer, but difficult to change. This technology can thus be transferred to other relationships with relative ease, but at the same time it cannot handle very much variation in the context of the exchange.

Tacit technology with a low degree of automation, flexible workflows, unspecified evaluations, complex untransferrable materials technology with much variation, a technology that handles many exceptions and that is difficult to analyse will have an interlocked activity structure that is difficult to transfer, but that can handle a great deal of variation in the context of the exchange. We cannot, however, find purely tacit relationship technologies in real life. Managing relationship technology would imply an understanding of the nature of this technology.

In volatile relationships and networks a tacit technology can handle variation, while a stable context can be handled by explicit technology. A technology with a low degree of tacitness is more difficult to change since the routinisation is bought at the cost of flexibility. A less tacit technology is easier to teach and transfer, but it is also a result of considerable investment in the form of automatisation, standardisation and routinisation. To change from a low degree of tacitness to a high one has its costs, mostly related to the learning of each firm's uniquenesses, and its surrounding network context. This is the reason why it is probably easier to go from a high to a low level of tacitness than the reverse.

The tacitness of the technology in one relationship has an effect on the relationships that are connected to it. For instance, a relationship with a technology that is not too tacit makes it fairly easy for the firm to describe the input from suppliers and the output to customers. It is probably often the case that this makes it easier for the firm to replace suppliers and customers. On the other hand, a firm with a tacit technology will probably have to develop a more complex understanding with its suppliers and customers. This understanding will probably require firms to commit more resources to match their needs and capabilities. To become involved in a

Kent Eriksson and Jukka Hohenthal

relationship with tacit technology demands more learning by doing and closer and more intense links between the parties to the exchange.

Consider two purely analytical cases, one in which all business relationships connected to one another have tacit technology, and another where all have a low degree of tacitness. These two cases may be considered to be a tacit network and an explicit network. The tacit network will be a system of tightly connected parties that frequently interact in order to understand how to adjust to one another. Among the actors, there is a strong need for an understanding of one another's business contexts, so that the complexities of the interlocked activity chains can be understood. This results in a closely knit system of interacting parties. Thus one actor alone cannot easily change work procedures because it requires him to fit in with the system. The explicit network is a much more open system where actors can change internal procedures as they wish, as long as they can specify the input and output needed. Those business relationships with explicit technology will be relatively easy to teach to other actors and so the technology can be transferred to another business relationship at a low cost.

The real world does not present us with purely tacit or explicit networks. Instead, there is a mixed degree of tacitness in relationships connected to one another. The aim of this paper has been to present a framework for understanding the relation between more and less tacit technologies in business relationships. The degree of tacitness in the technology of the relationship is seen as a key to understanding matters such as the exchangeability of business partners and the ease of changing technology. For instance, a business relationship between two parties will be more closely related to their network context if adaptations have a high degree of tacitness. In such cases, the business relationship and the business network should be studied as one entity. A different kind of study is called for if adaptations have a low degree of tacitness. In this case studies can focus on the relation between network and relationship, or on their influence on each other.

References

ALTER, C., and HAGE, J., 1993, *Organizations Working Together.* Newbury Park: Sage.

ANDERSON, E., and WEITZ, B., 1992, The Use of Pledges to Build and Sustain Commitment in Distribution Channels *Journal of Marketing Research,* Vol. XXIX, pp. 18–34.

ANDERSON, JAMES C., HÅKANSSON, HÅKAN and JOHANSON, Jan, 1994, Dyadic Business Relationships Within a Business Network Context *Journal of Marketing,* 58 (October), pp. 1–15.

BARLEY, S .R., 1986, Technology as an Occasion for Structuring: Evidence from Observations of CT Scanners and the Social Order of Radiology Departments *Administrative Science Quarterly,* 31, March, pp. 78–108.

BARLEY, S. R., 1990, The Alignment of Technology and Structure through Roles and Networks, *Administrative Science Quarterly,* 35, March, pp. 61–103.

COLLIS, D., 1996, Organizational Capability as a Source of Profit in Moingeon, B., and Edmondson, A., (eds) *Organizational Learning and Competitive Advantage*, pp. 139–163, London: Sage.

ENGWALL, L., 1978, *Newspapers as Organizations*, Farnborough: Teakfield.

FORD, D. (ed.), 1990, *Understanding Business Markets: Interaction, Relationships and Networks*, San Diego: Academic Press.

HÅKANSSON, H., and SNEHOTA, I., (eds) 1995, *Developing Relationships in Business Networks*, London: Routledge.

HALLÉN, L., JOHANSON, J., and SEYED-MOHAMED, N., 1991, Interfirm Adaptation in Business Relationships *Journal of Marketing*, 55, April, pp. 29–37.

HUTCHINS, E., 1991, Organizing Work by Adaptation *Organization Science* Vol. 2, No. 1 pp. 14–29.

LAWRENCE, P. L., and LORSCH, J., 1967, *Organizations and Environment*, Boston, MA: Harvard University Press.

LEVINTHAL, D. A., and FICHMAN, M., 1988, Dynamics of Interorganizational Attachments: Auditor-Client Relationships *Administrative Science Quarterly*, 33, pp. 345–369.

LEVITT, B., and MARCH, J. G., 1988, Organizational Learning *Annual Review of Sociology*, Vol. 14, pp. 319–340.

MARCH, J. G., 1994, *A primer on decisionmaking*, New York: Free Press.

MAYER, ROGER C., DAVIS, JAMES H., and SCHOORMAN, F. David, 1995, An Integrative Model of Organizational Trust *Academy of Management Review*, 20 (3), pp. 709–734.

MORGAN, R .M., and HUNT, S.D., 1994, The Commitment-Trust Theory of Relationship Marketing *Journal of Marketing*, 58 (July), pp. 20–38.

NOHRIA, N., and ECCLES, R. G., (eds), 1992, *Networks and Organizations: Structure, Form, and Action*, Boston, MA: Harvard Business School Press.

NONAKA, I., 1991, The Knowledge – Creating Company *Harvard Business Review*, November/December, pp. 96–104.

NONAKA, I., and TAKEUCHI, H., 1995, *The Knowledge – Creating Company*, New York: Oxford University Press.

POLANYI, M., 1969, *Knowing and being*, Chicago: The Chicago University Press.

PORAC, J., THOMAS, H., and BADEN-FULLER, H., 1989, Competitive groups as cognitive communities: The case of Scottish knitwear manufacturers *Journal of Management Studies*, 26, pp. 397–416.

POWELL, WALTER W., KOPUT, KENNETH W., and SMITH-DOERR, LAUREL, 1996, Interorganizational Collaboration and the Locus of Innovation: Networks of Learning in Biotechnology *Administrative Science Quarterly* 41: pp. 116–145.

RING, PETER S., and VAN DE VEN, ANDREW H., 1992, Structuring Cooperative Relationships between Organizations, *Strategic Management Journal*, 13, pp. 483–498.

RING, PETER S., and VAN DE VEN, ANDREW H., 1994, Developmental Processes of Cooperative Interorganizational Relationships *Academy of Management Review*, 19, January, pp. 90–118.

SCOTT, W. R., 1981, *Organizations*, Englewood Cliffs, NJ: Prentice Hall.

SJÖBERG, U., 1996, *The Process of Product Quality Change Influences and Sources: A Case from the Paper and Paper-related Industries* diss. Department of Business Studies, Uppsala University.

TYRE, M. J., and VON HIPPEL, E., 1997, The Situated Nature of Adaptive Learning in Organizations, *Organization Science*, Vol. 8, No. 1, pp. 71–83.

WALSH, J. P., and UNGSON, G. R., 1991, Organizational Memory, *Academy of Management Review*, Vol. 16, No. 1, pp. 57–91.

WALUSZEWSKI, A., 1989, Framväxten av en ny mekanisk massateknik. diss. Department of Business Studies, Uppsala University, Uppsala: Acta no 31.

Kent Eriksson and Jukka Hohenthal

WEICK, K. E., 1979, 2nd ed., *The social psychology of organizing* Reading, Mass.: Addison-Wesley Publishing Company.

WEICK, K. E., 1990, Technology as Equivoque: Sensemaking in New Technologies, in Goodman, P. S., and Sproull, L. S., 1990, *Technology and Organizations*, pp. 1–44: San Francisco, CA: Jossey-Bass.

WEICK, K. E., and ROBERTS, K.H., 1993, Collective Minds in Organizations: Heedful Interrelating on Flight Decks, *Administrative Science Quarterly*, Vol. 38, No. 3, pp. 357–381.

WILLIAMSON, O. E., 1975, *Markets and Hierarchies: Analysis and Antitrust Implications* New York: The Free Press.

WINTER, S. G., 1987, *Knowledge and Competence as Strategic Assets*, in Teece, D., *The Competitive Challenge. Strategies for Industrial Innovation and Renewal*, pp. 159–185: Ballenger. Cambridge, MA,.

WOODWARD, J., 1965, *Industrial Organizations*, London: Oxford University Press.

ZANDER, U., 1991, *Exploiting a Technological Edge – Voluntary and Involuntary Dissemination of Technology*, IIB, Stockholm School of Economics.

CHAPTER 7

Collective Innovation – The Case of Scania-Cummins

LARS SILVER and TORKEL WEDIN

Introduction

Within the industry for trucks with a gross vehicle weight (GVW) over 16 tonnes (as for the entire automotive industry) there is a constant need to produce more efficient vehicles, achieving higher performance. This is especially important in terms of fuel consumption as this is the largest individual operating expense for vehicle owners, and therefore is an important variable in the transport economy. As the price of fuel increases, it becomes even more important to reduce fuel consumption. Another external force is environmental organisations that continuously put pressure on the truck industry to decrease the impact trucks have on the environment. If an engine designer is to reduce fuel consumption, one of the most critical parts of a truck is the fuel injection system. In many respects, this system also determines the performance capacity of other parts of the engine.

The focal company in this case, Scania, is one of the world's leading producers of vehicles over 16 tonnes. For a number of years the fuel injection system was purchased from Bosch, a German producer that totally dominates this industry. However, during the last eight years Scania has started a new relationship with an American engine manufacturer, Cummins. This case illuminates the possible reasons why this shift in relationships occurred and how the technology used influences the choice of relationship partners.

Lars Silver and Torkel Wedin

Points of departure

Studies on technological development processes are often based on stories of internal development processes within a given firm. Furthermore, most of these studies present "success stories" where the individual firm manages to develop new products on its own. One of the most common notions is thus that a producing firm initiates and finalises technological developments without having to take into account the effects of external actors. The external environment is considered to have little or no impact on development processes (Utterback and Abernathy 1975).

An alternative way for a firm to develop internally is to buy the technology on the market. The underlying assumption of inter-firm relationships in these cases is often a market view, where exit (and therefore also the price mechanism) is described as the only way to communicate with suppliers and customers.

There are, however, several scholars that provide empirical evidence in contradiction to the aforementioned statements. More specifically, they argue that in many cases product development is the result of interaction between suppliers and customers (among those arguing along these lines are von Hippel, 1988; Håkansson, 1989; Anderson *et al.* 1994).

It is possible to discern three different types of governance modes regarding the organisation or co-ordination of innovation. The first is when an innovation is seen as an internal affair and is governed by the internal hierarchy. However, if the firm is not able to develop the technology itself, there is always the possibility of buying. The development mechanisms are then governed by the market. The third type of governance mode, the relationship or network type, stresses co-operation in product development, instead of relying on the price mechanism or the internally controlled innovation efforts. One could wonder why firms organise themselves in these types of governance modes? Given that firms act under "norms of rationality" there are good reasons to believe that different co-ordination forms for developing, using and selling new technologies are conscious choices. However, the missing link in this way of arguing so far is in our opinion the building blocks of firms and markets, namely the inherent resources and their properties.

Resources, learning and innovation

Penrose (1959) describes the firm as a set of productive, heterogeneous resources. Every firm is unique which is explained by the difference between the resources and the services these render. In her terms, the task for the firm is to organise the use of the resources controlled by the firm, together with resources purchased from the firm's external environment. Resources have meaning only when regarded as constellations, when they are combined with other resources in order to generate known useful outcomes (Håkansson and Snehota 1995, p. 133). Resources are tied together in such a way that one tie will inevitably influence other ties within the same constellation of resources. The resource constellation reflects the knowledge of resource use, or technology, in the business network. In order to develop the resource constellation, there is a need to co-ordinate the learning process. Learning within a network can take the form of experimentation by one single actor. It can also be a matter of transferring knowledge between two actors, or as a process of joint learning, where several actors learn more or less simultaneously from the experiences of other actors. In the learning process, a single actor can try to acquire exactly the same knowledge that other actors have. This is rather uneconomical, though, and therefore it is often wiser to specialise for the mutual benefit of all actors (Demsetz 1988, p. 157).

Resources are being tied together and thus combined in innumerable ways, which leads to different outcomes and to different economic values. This is why resources are said to be heterogeneous (Alchian and Demsetz 1972). There is always a potential for improvement in the application of resources, either within the confines of a particular firm, or across the borders of a firm. In learning about the heterogeneity of resources, it is possible to achieve technological innovation and network change. The notion of resource heterogeneity is closely connected to learning and knowledge. Since an actor can never know all the possible combinations in which resources can be tied, because of cognitive limitations or merely because the available technology is inadequate, it is of interest to continuously advance the knowledge of resource combinations.

Resources develop over time as adaptation and refinement in the application of resources make a resource blend into its environment, thus virtually erasing the boundaries between resources. But not all resources are blended into the environment. Or at least, they are not uniquely adapted to one set of circumstances. In some industrial relationships, resources are virtually inseparable. In other relationships resources have been standardised to fit into different environments. The ties of resources will differ depending on the importance of these ties, as perceived by the actors. The ties of resources and how they are utilised are also dependent on the costs of developing these ties and the competence that actors posses, that is their ability to utilise them. These variables determine how much effort goes into the development of certain resources and resource constellations.

The issue that this chapter raises concerns whether internal development, or co-operation in a market, is a conscious choice made by the actors, or if it is the resources tied in complex technological systems that determine the modes of development processes. The focus of this chapter is to explore the impact of different kinds of resource ties between companies, and if so, how this influences strategies for product development and, therefore also, how learning takes place. The empirical evidence supporting this line of argument is set in the above mentioned truck industry, and focuses on the development of a new fuel injection system. The case has been chosen because it highlights the effects of technology on product development strategy.

The case – the development of a new fuel injection system

In this section the empirical case, a development project between Scania and Cummins, is presented. The case is based upon semi-structured in-depth interviews with those in both the top and middle management at Scania who were involved in the development project with Cummins[1]. To gain a deep understanding of the truck industry, the case is also supported with technical literature about engines in general and fuel injection systems in particular.

A truck over 16 tonnes needs a powerful engine to enable it to perform its tasks adequately. Trucks are comprised of different systems that are fitted together, one of which is the engine. In its turn, the engine – as a technical system – can be said to consist of a number of subsystems. The basic engine consists, for example, of pistons, connecting rods, cylinder heads, etcetera. To these components others are fitted: the supercharger (where air is "pushed in"), for example, which can be considered to be a system in itself. Oil refining systems are another example, combining the oil system with a refinement system.

One subsystem amongst all these subsystems in the engine is the fuel injection system, a system closely integrated with the engine. The fuel injection system is the core system in many respects, influencing many of the properties of an engine and therefore largely determining the engine's capabilities, such as the engine's fuel consumption, manoeuvrability, engine characteristics, overall effectiveness and environmental aspects. The provision of this fuel injection system is currently recognised as one of the most high-tech tasks in the truck industry.

The function of fuel injection systems is to deliver the right amount of fuel and air to the cylinders. Even if this function is fairly easy, the manufacture of the fuel injection system requires far higher precision than any other system in the vehicle. Furthermore, since the system itself is mechanically quite simple, a controlling device able to estimate the optimal fuel

---

[1] The case is developed from an earlier study by Adermalm, Sjöberg & Wedin (1998).

allocation under each set of circumstances is needed to influence the fuel consumption of the vehicle. The more precise the controlling system is, the more efficient the fuel consumption will be. Thus, to further enhance the properties of diesel engines, thereby yielding a reduction in fuel consumption and exhaust-gas emissions, the manufacturers of fuel injection systems and their customers have to rely on the manufacturers' ability to develop accurate start-of-injection timing, precision-manufactured injection nozzles, precisely-defined fuel-spray geometry, and further increases in injection pressure. The fuel injection system has its major interfaces with the pistons and the turbocharger. Together, these are the most important components influencing the performance of the truck engine.

Since the 1980s, technological improvements have meant that electronics has become a vital factor in the design of engines. Gradually this has also sharpened the demands made on the producers of fuel injection systems to adapt their products to the overall requirements of the electronics around the systems. In particular, this development has been hastened by increased pressure from EU authorities and governments to protect the environment from pollution. The growing importance of electronics has resulted in the increased use of development contracts. In part, this is due to the fact that development costs have been harder to assess.

Scania saw this development as a chance to further improve its engines. In doing this they needed help from a supplier of fuel injection systems. The most logical solution was for Scania to ask Bosch, the company with which they had worked for a very long time to develop a new kind of electronic fuel injection system. However, Bosch turned Scania down, or, more precisely, suggested another solution. Scania was not satisfied with the solution, which they perceived to be a minor incremental upgrade of their previous system, so they started to look for another development partner.

Presentation of the actors

Scania

Scania was founded at the beginning of the century when the two companies Vabis and Scania merged. Scania has about 24 000 employees and the sales for 1997 were 40 billion SEK. Scania is a public company and its biggest owner is Investor, the largest investment company in Sweden.

Historically Scania has focused on the development and production of trucks and buses over 16 tonnes. Scania is the third largest manufacturer of heavy trucks in Europe, after Mercedes-Benz and Volvo, and in 1994 Scania's production amounted to some 31 000 trucks in total. Worldwide, Scania is the number five producer in the heavy truck segment. Competition within the industry is fierce and there is continuous pressure on manufacturers to reduce production costs. Since the mid-1970s, the truck manufacturing industry has changed dramatically in Europe. Many large producers have been acquired by even larger competitors. Scania has remained independent, and has not been engaged in acquisitions.

Scania currently sells in every major market except North America. The production plants are located all over Europe and in South America. In 1985 and 1986 Scania started to sell its product lines on the North American market. As the North American market is equal to a third of the world market, this was an obvious choice for expansion. Unfortunately, the sales remained weak despite ongoing efforts to coordinate production with Ford. Ultimately it was decided that due to the particular demands of American customers, authorities, traditions, etc., the only component that could be used from the original product line was the engine. In part, this can be explained by the fact that American manufacturers more often assemble pre-fabricated truck parts, while European manufacturers have a large amount of inhouse production. All other components have to be custom-made for the North American market. Scania soon abandoned its attempt to gain a foothold in the American market. The same can be said for the Japanese market, which remains another of the few blank areas on Scania's sales map.

It is estimated that there are 350 000 Scania trucks in service around the world, which means that over half of the Scania trucks ever produced are still running. The engines in the trucks that Scania produces have to be able to withstand the damage incurred by running 2 500 000 miles without major overhauls. Obviously, this demands that Scania engines are extremely reliable and durable. Scania's truck production is noted for its high degree of specialisation: they do not manufacture smaller trucks at all. This is because it is relatively hard to use the same components for a light truck and a heavy truck to obtain economies of scale. Indeed Scania has argued that there are virtually no economies of scale at all: every component is different: engines, gear boxes, chassis, etc. Furthermore, Scania's production is characterised by a high degree of vertical integration and it has a highly modularised production system. During the 1960s Scania decided to produce the principal components of the trucks inhouse. Unfortunately, this also meant that some economies of scale might have been lost. At the same time Scania has effectively managed to reduce the number of components needed for their production to some 20 000, which is far less than their major competitors. Scania will go to great lengths in their attempts to reduce the total number of components needed to keep their engines running. Only some 350 of the components are located in the diesel engine, in total, the engine and the transmission system requires some 3 500 components.

Scania currently has some 800 suppliers that provide the components needed to manufacture an engine. Some 50% of the components are designed inhouse, while the suppliers design the remainder. It is rather common for truck producers in general to engage in joint development products with suppliers, both in the form of continuous, long-term research projects, and shorter, more focused projects. The trend is also that these kind of joint development projects become more important as electronic technology gradually supersedes earlier mechanical solutions.

As research becomes more complicated, it is necessary to maintain close relationships and development projects with suppliers. Collaboration requires time and investment, and the choice of which supplier to work with becomes increasingly important. Primarily, the development still takes place inhouse, which means that setbacks in the development process can be corrected within each firm. The competitiveness of Scania to a large extent originates in their knowledge of engine development and engine production. Scania has chosen to maintain its profile as a "developing" firm, instead of becoming a "buying" firm. The consequence of this is that Scania routinely retains five or six development projects solely for the purpose of developing engines.

Scania's policy regarding its suppliers has always been to maintain at least two separate suppliers for crucial components and systems. Despite this, Scania had only a single supplier for the fuel injection system during the 1970s and 1980s. This supplier, Bosch, was and still is one of the leading suppliers worldwide for the automobile industry and has dominated the market for these types of systems. Consequently Scania has had a long relationship with Bosch and has been one of its customers since the mid-1950s. The relationship with Bosch was and still is perceived as a positive and well functioning one within Scania.

Bosch

Bosch is a German company situated in the southern part of Germany. The company's total turnover is about DM 35 billion, with the automotive area accounting for more than 50%. Of the ten divisions within the automotive area, the one concerned with fuel injection systems for diesel engines is the largest. Bosch is the world's largest independent manufacturer of automotive components and has a large market share: about 70% in the segment comprised of trucks over six tons. Bosch's product development mainly takes place in Stuttgart in Southern Germany. There is, however, a trend in Bosch towards moving product development abroad since the company is getting involved in more and more projects outside Germany (Wynstra 1997).

Bosch works together with German manufacturers, like Mercedes Benz and BMW, the Dutch DAF Trucks and Scania. Bosch also has close contacts with producers of pistons, like Kolbenschmidt and Mahle from Germany. This is due to the interface between the fuel injection system and the pistons (Wynstra 1997).

Bosch has, as mentioned above, been a supplier to Scania since the mid-1950s. During the 1980s, Scania was seen by Bosch as a pilot customer and was also the "lead user" when the electronic in-line pump system (EDC) was developed and set in production in 1987.

Towards the end of the 1980s, there was something of a revolution within the area of fuel injection systems when the pump injectors and electronic injection systems were developed. This change can be considered to be a technological shift, which provided the opportunity to develop something new to replace the previously dominating mechanical fuel injectors. Up to this point Scania had purchased the fuel injection system from Bosch alone. Scania perceived during the late 1980s that Bosch lacked the willingness and drive to advance technology as fast as Scania wanted. This was particularly so in the development of electronically controlled pump injectors. One of the reasons for this could have been the investment Bosch had made in equipment for in-line pump production, which may have taken away the incentive for any advances in injection pump technology. Even though the relationship with Bosch functioned very well on the whole, Scania was looking for a complementary supplier. Scania realised that there was an opportunity to decrease the dependence on Bosch as the sole supplier, and they systematically investigated the possibilities of using alternative suppliers.

Cummins

In 1988, with the intention of finding a second supplier of fuel injection systems, Scania approached a number of American truck engine manufacturers and started to investigate the possibility of establishing a long-term relationship for the development of a new fuel injection system. Scania's hopes were for a considerable reduction in the fuel consumption, and that the company would be able to meet the higher environmental standards expected. The manufacturers that Scania approached were Caterpillar, Detroit Diesel, Navistar, and Cummins. In the end, Scania found that Cummins seemed to have consistent ambitions and therefore the two companies chose to establish a joint venture. This did not mean that Scania ceased working with Bosch, but rather, that Scania's efforts to advance technology in the fuel injection field were primarily directed towards the relationship with Cummins, instead of with Bosch.

The US based firm Cummins is one of the largest manufacturers of diesel engines in the world. In 1990, engines amounted to 69% of its sales.

The firm has a complete product line of engines ranging from 76 to 2000 horsepower, which are used by all kinds of customers. The principal market is the North American heavy-duty truck industry, where every major truck manufacturer offers Cummins' engines as the standard or the option. The market share was 46% in 1990. Cummins is said to be a decentralised firm, where each manager can act independently (Laage-Hellman 1997).

Cummins Engine has had some prior experience of elaborate joint development projects. One project was a joint development with Toshiba on the use of ceramic components in diesel engines. This project eventually led to the formation of a joint venture firm located near Cummins' headquarters in the US. The aim was to sell the products developed originally for the Japanese market to North American truck manufacturers. In fact, the joint venture firm ended up selling engines to some of Cummins' main competitors in the US: Caterpillar, Detroit Diesel and Mack Trucks. The relationship between Cummins and Toshiba was quite intimate and involved a large number of employees on both sides. There were also numerous channels of communication available and information flowed at different hierarchical levels. The development of ceramics has not yet been a commercial success in the short term. However, both Cummins and Toshiba (especially the latter) have deemed it to be a strategic material of the future. Thus, both actors remain convinced that some sort of joint development programme needs to be maintained. Even though the relationship between Cummins and Toshiba remains cordial to this day, Cummins has not been hesitant about looking for other solutions simultaneously, some of them involving Toshiba's competitors in the US. Cummins usually works on joint development projects with different actors simultaneously, a trait they share with most actors in the field, including Scania and Bosch.

As a truck engine manufacturer, Cummins was a competitor to Scania. So why would Scania choose to collaborate with Cummins under these circumstances? One reason was the fact that the distance to the North American market was prohibitive, and Scania had never been one of the major actors in this market. Furthermore, Cummins had no legal ties whatsoever with other actors that could be regarded as competitors to Scania, therefore the relationship was perceived as "safe" for Scania.

Organising the relationship between Scania and Cummins

In 1989, Scania embarked upon a joint development project with Cummins, the largest diesel truck engine manufacturer in the world. To start off, the partners got to know each other and the first project commenced with fuel injection systems. It was not until 1992 that the first formal contract was signed and the relationship was finally formalised as a joint venture around a so called "high pressure fuel injection system". Before this signing, some development work had been conducted and both of the parties felt that a continued collaboration agreement would prove worthwhile. The original development contract was seen as the "real" start in the development of a new generation of fuel injection systems.

The relationship with Cummins was perceived as somewhat tense in the early stages by some of the people within Scania. The staff involved at Scania asked themselves whether they could conceivably share their knowledge with the staff of Cummins. The representatives of Cummins were quite candid and open in their relationship with Scania employees, a surprising attitude according to some Scania officials who, perhaps, had expected more of a John-Wayne attitude from the American company.

Communication between the firms was mainly in the form of daily telephone contact during the development project. In some periods of hectic work there were also videoconferences once a month. It was also necessary for staff members from each side to get to meet one another face-to-face to discuss the ongoing development projects. For these purposes there were four "steering committee meetings" held each year in either Sweden or the United States. During these meetings the project was evaluated and measured against the pre-planned schedule of development. During the meetings the philosophies of each firm were linked together, and a mutual understanding was reached between the firms.

The steering meetings were organised according to the following schedule: during the first day a technical meeting was held, a meeting where the technicians presented and discussed the progress and shortcomings up to date. The next day there was a steering committee meeting at which the technicians were joined by leading staff members from both firms. On the third day there was a meeting with the chairmen at which the CEOs of the firms participated. All in all there were about 20 steering meetings from the day the project started until the project was officially terminated.

For the staff involved the project was very demanding and challenging. One problematic aspect was the physical distance, but the separate time zones further complicated the collaboration process. The communication between staff members in the two organisations was significantly enhanced when the memo-system was installed, and later on when it became possible to send documents by e-mail.

In order to decrease the physical as well as psychic distance, a key individual in Scania was sent for a long-term visit to Cummins, in order to participate in the development project. This visit lasted from 1992 to 1996, during which time the individual worked in the Cummins fuel injection group. Furthermore, the visitor was responsible for the development and testing process, and had five Cummins employees as subordinates in the testing department. When the visit ended, Scania engines were on site at the Cummins laboratories, enabling continued experimentation. Today there are four Scania engines and one whole truck in the Cummins laboratories in different test-cells. Scania also sent a replacement for the individual in charge of the testing programme. Scania have also received a visitor from Cummins who works on Scania's own electronic control system. This collaboration programme is also a very important part of the relationships.

Scania have had a couple of work groups associated with the Cummins relationship over a number of years. In the United States, the number of Cummins' employees working on the Scania project is even larger, possibly some 50 staff members.

Problem solving activities in the relationship

During the development project a large number of day-to-day problems and more strategic problems arose that had to be solved. One such problem was how a fuel injection system is actually constructed and what parts it consists of. On a general level, every fuel injection system consists of mechanical and electronic components[2]. Since this requires knowledge in two widely separated fields, the companies had to find staff members capable of solving technological problems of both mechanical and electronic origin. There were groups working on the mechanics as well as groups working on electronics in both Cummins and Scania during the development phase. The two disciplines tried to link their respective competencies to find common solutions. Within the project there were some common difficulties, and when these difficulties emerged in test engines or test vehicles it was crucial to have both groups in the vicinity of the testing site.

Another problem was the ambition to standardise components, which was in conflict with the perpetual need to introduce changes in some way. The reason is that every fuel injection system has to be carefully adapted to each engine. Scania's ambition is still to strive for standardisation whenever possible and they regard themselves as being quite successful in standardising equipment and components. At the same time, there are

---

[2] The electronic control system communicates with all the different systems in the vehicle. A large number of systems need the information provided by the engine. A prime example is Scania's Opticruise system, in which the systems are totally integrated.

always some aspects that differ between the engines when it comes to the demands placed on fuel injection techniques[3].

In most development projects problems arise that threaten the entire project. This case was no exception. In 1993 tough problems were encountered, and the future of the project was in considerable doubt. Originally Cummins had a system that principally controlled the pressure in the fuel injection, which used a choke. With higher pressure, more fuel was injected into the engine and the driver controlled the injection process through the throttle. When the driver let go of the throttle, the pressure was minimised and the engine idled. There were, however, engine response problems that had to be solved. This came about because the throttle failed to respond instantaneously, so the project was forced into using a different approach. The project strove for a solution with magnetic valves, where the valve was open for ten milliseconds, thus generating a pulse motion providing a fast and exact fuel pressure.

The fuel injection system had to be totally restructured. Instead of controlling the pressure, it was now to be pulsated into the injector under constant pressure. The discussions regarding this solution were fierce, and the parties saw this as something approaching a complete re-start for the entire project. Because of the decision to restructure the entire system, Scania had to rebuild their engines. Cummins, however, elected to go on with the old system, but only with their largest engines made for industrial and marine purposes. The original idea had been to start production in 1995. This aim had to be abandoned when the re-start occurred in 1993.

Scania arguably provided the crucial motivation for Cummins to continue with the project when difficulties in the development process emerged, in the same sense that Cummins acted as a motivation for Scania to continue working on the project. Occasionally, the difficulties encountered were severe enough to motivate an early end to the project, and there were individuals who argued for closing it down. Yet, as there were two parties in the relationship, it was hard for either Scania or Cummins to defend a decision to end the project without a similar decision being reached by the other party.

---

[3] For example, the lower parts of the fuel pumps are not standardised as the gaps need to be individually fitted for each engine. The upper parts are not all standardised either. This has to do with the overhead block. Cummins has an overhead cam shaft in their cylinder head, and they have a large cylinder head altogether, which Scania does not use. This is a sign of different construction philosophies. Scania wants to have one cylinder head per cylinder, which prevents the use of an overhead cam shaft, instead they have a common cam shaft, which in turn makes the overhead construction different. Scania utilises an overhead block, while Cummins chooses not to do so. Otherwise the components are similar in the solutions used by the two firms.

How Cummins' network affects the relationship with Scania

The contract Scania have with Cummins stipulates that Cummins cannot sell the system to other truck manufacturers without prior agreement from Scania. This could, however, pose a problem for Scania in the future. Scania is well aware of the fact that Cummins is the largest diesel engine manufacturer in the world, and that it surely wants to use its acquired knowledge from the project to sell the newly developed electronic fuel injection system to other truck manufacturers. How the firms will solve this problem is as yet unclear even if there are clauses in the contract stipulating how Cummins may use the system. Thus, it is hard to envisage how competitors should not be able to purchase this motor from Cummins. This would eventually mean that other truck manufacturers would have access to the same fuel injection system in the engines sold in the United States that Scania has.

Cummins is currently engaged in development projects with several actors in the heavy-duty vehicle industry, primarily in conjunction with actors in the North American market. One interesting deviation from this is their development projects with Bosch for developing propulsion equipment. The reason for these kinds of development projects is that they do not perceive each other to be competitors in this field. Other fuel injection system manufacturers, such as CAV, Bosch and Lucas are not engine manufacturers. Instead they only manufacture the fuel injection system, even if they also manufacture a number of other components and systems for the vehicle industry. Cummins is, however, an engine manufacturer that also manufactures fuel injection systems. This means that it has a wider knowledge that can be exploited in its relationship with truck manufacturers, because its competence centres on the whole engine rather than just the fuel injection system.

Cummins have always manufactured their own fuel injection equipment for their large engines. In conjunction with this they also manufacture smaller engines for pick-ups and smaller buses. In these cases Cummins uses the Bosch system and purchases from Bosch just like Scania does. Cummins co-operates with truck manufacturers in that it sells engines to them and it has collaborative projects ongoing with other engine manufacturers as well. These are entirely focused on larger engine manufacturers in terms of high horsepower engines.

The role of knowledge in the relationship

The aim of the collaboration from Scania's side was and still is to gain access to know-how and to develop its own knowledge. While Cummins possesses knowledge about pump injectors and electronic control systems, Scania can offer knowledge about systems in the drive train including gearbox and retards. These different parts have to be integrated when developing modern drive train systems. Thus, the knowledge possessed by the two firms was complementary, but also overlapped to some extent. Even if the development project demands a substantial amount of money without any prospect of an immediate return, the project is still pursued in order to guarantee a certain level of future competitiveness and to ensure long-term survival. Scania claims to have gained knowledge in a way that would otherwise have been impossible, and this knowledge can then be transferred to other application areas.

Since the time that the development project started, it has been developed and deepened, until today it also incorporates the development of systems other than the original fuel injection system – one such example being the injector drive train in the engines manufactured by Cummins.

The injector drive train is the same for the earlier Bosch system as for the Cummins system, and Scania have improved their injector drive train to be compatible with the Bosch system too. In this case, Bosch lacks the knowledge of how engines have to be constructed to take the additional strain caused by moving the fuel injection system to the top of the cylinder. In this field Cummins is, of course, highly capable. So, when Scania installed their Bosch injector, they had a significant amount of Cummins' technology in their cylinders.

Scania's aim is to have a holistic view of vehicle construction and systems, such as, transmission systems, brake systems, the functions behind the steering, gearboxes and so forth. In this area Scania have been able to impose requirements that the systems Cummins produces must be compatible with their components in different areas. In this way, Cummins has learnt how Scania uses the engine control system for different purposes in the chassis construction process. Scania have been able to teach Cummins about the requirements of vehicles as a whole, instead of merely the requirement of the engine. While Cummins is highly capable of adapting engines to meet various needs, they possess less knowledge about the needs different vehicles have. Cummins has also been able to use the knowledge that has been created and developed in the relationship with Scania in the North American market, thanks to advice given by Scania.

The outsourcing of the vital parts in the engine could negatively affect the future competitiveness of a firm in this business. It could prove worthwhile in the short term because it forces suppliers to standardise their products, but in the end, the knowledge about these systems will disappear. In the case of Scania, it is vital for them to retain that system knowledge if they wish to remain a "driver" of technological change in the field. Even if the components can be purchased, an understanding of the overall system cannot. Only by participating in the development of all the systems and components in the engine is it still possible for Scania to seriously influence the future of the truck industry. Without their unique knowledge, they will cease to be an attractive partner for development projects. In the long run this could make the difference between being a first-rate player actively engaging in the development of the truck industry, and being a second-rate manufacturer only reacting to initiatives made by other manufacturers.

By not developing and producing these systems alone, but together with others, it could be argued that Scania is making itself dependent on specific suppliers. So, should Scania instead manufacture these types of systems itself to stay independent and is this strategy viable? It could be argued that electronic fuel injection systems are more or less a totally new technology for Scania, which is a definite drawback. In manufacturing fuel injection equipment, the tolerances are 1000s of millimetres, and the whole array of manufacturing equipment required for this production is different from that with which Scania has experience. Scania has not really considered this step so far, and the investments in capital and know how necessary for this enterprise are staggering; instead Scania is striving to improve its relationships with its suppliers. The dependence is not one-way, however. Firms such as Cummins and Bosch are dependent, in their turn, on engine and vehicle manufacturers such as Scania because they can act as conveyors of market information and customer requirements.

Case epilogue

In the beginning of 1999 Scania and Cummins announced the partnership officially.

> *"The new high-pressure fuel injection system jointly developed by Scania and Cummins is now entering the production phase. The system is designed to enhance efficiency and environmental performance" (Scania Press release 7 January 1999).*

All this to meet the new, stricter environmental standards, as well as the competition from alternative systems provided by other suppliers. Scania and Cummins formed a joint venture from 1st January 1999 – Cummins-Scania High Pressure Injection Inc LLC, with production in Columbus, Indiana. The joint venture is 70 percent owned by Cummins, and Scania owns 30 percent. The first engine to use the system is a Cummins' engine, Signature 600. Both parties seem to be happy with the alliance. "The R&D period has been very constructive", says Scania's Chief Technical Officer. The CEO of Cummins, Jim Henderson, is also happy: "We are pleased to be able to further our relationship with Scania, with whom we developed this exciting new technology" (Scania Press release 7 January 1999).

But one might wonder what happened with Bosch, then? Is the German company hopelessly lost behind Scania and Cummins? No, not at all. In fact, soon after Scania embarked upon its relationship with Cummins,

Bosch started to develop electronic pump injectors for a new fuel injection system, and it beat the Swedish-American alliance to the market.

Apart from experimental products, Scania makes no direct purchases from Cummins today. Instead, all equipment used in the fuel injection systems at present is bought directly from Scania's old supplier, Bosch. In the future, however, Scania hopes to allocate 50% of the purchases to each company.

Analysis – the economics of inter-organisational learning

From the case, we have been able to distinguish three different types of resource interfaces that affect firm behaviour.

First, we can see that the old type of fuel injection system was a type of resource, controlled by Bosch, that the other firms were more or less *dependent* upon. The resource interfaces of other resources were adapted to this resource, the fuel injection system, and not the other away around. Bosch wanted things to stay this way. Therefore there was no need to engage in a joint project with Scania. At the same time there were few possibilities for Scania in this case to influence the development of the fuel injection system as long as Bosch was the only counterpart in terms of this product. One can for example argue that it was only when Scania decided to start working with Cummins that Bosch saw an incentive to develop a new type of fuel injection system.

Second, in terms of learning and knowledge, the case emphasises that knowledge in itself provides little value, it has to be used or combined with complementary knowledge if it is to provide any value for those possessing it. Therefore, knowledge needs to be complementary since the actors must receive something they themselves lack, but it needs also to be overlapping to the extent that they need to have a common framework to act within. Compared to the example above, the resource interfaces in this case can be said to be *interdependent*. Knowledge is not only embedded in the minds of humans, but also in technical equipment, which in itself remains capital intensive. The investments in the "hardware" part of knowledge lead actors in some certain directions. Furthermore, the choice of partner with whom to develop knowledge is crucial. And yet, the choice is made within a certain framework: the partner cannot be just anyone, instead it has to be an actor that is connected technologically to the company in one way or another. To find a partner that the company can gain new knowledge from, the company must have something to offer in exchange. This means that the history of the company becomes important. To be able to "give" something today, one must have received something in the past. A company, then, cannot be "peripheral", but must be more or less in the mainstream both technologically and economically. In this way, a partner can relate to a specific kind of knowledge, claim it for its own, and then transform it, to give it back to its counterpart in the next step.

Third, the case also shows that knowledge generated in one relationship must be of such a nature that it can be transferred for the actors to gain the benefits of co-operation. If this is not possible, product development and knowledge generation becomes virtually meaningless: knowledge and learning has a significant cost side that has to be taken in consideration. Without the possibility of eventually spreading the costs, even to actors not involved in the actual development process, no actor would engage in product development activities. This is something that both actors will benefit from and have to be aware of. There is a substantial difference between a truck manufacturer such as Scania, and the American truck manufacturers, who in reality are merely assemblers. Scania has to accept the costs of product development, including all the failed projects necessary to identify the projects that can be transformed into capital generating ideas. The American assemblers can avoid these costs and focus on becoming more efficient in their "manufacturing" of trucks. However, Scania has one advantage over its American counterparts: it has the ability to influence the product development agenda, so it can act, while American truck manufacturers merely react. Whether the strategy used by Scania is more beneficial in the long run is hard to envision, although it is quite obvious that Cummins needs an active partner such as Scania to advance its technology. At the same time, both Scania and Cummins need the American manufacturers to indirectly share the costs of development. Some sort of standard must emerge, the resource interfaces must be *independent*, so the actors involved can reach some economies of scale and get some pay off from the R&D investment.

In this chapter we have posed the question: does the behaviour of a firm and the selection of its counterparts have anything to do with the different types of resource ties that exist between the firms? Above we identified three kinds of resource interfaces:

1. Dependent resource interfaces;
2. Interdependent resources interfaces;
3. Independent resource interfaces.

The implications that can be traced from the different types of resource interfaces are shown below.

Implications of different types of resource interfaces for organising and learning

Dependent resource interfaces exist when a system of interconnected resources is arranged around, and adapted for, one single critical resource, then every other resource within that system can be said to be dependent on this critical or dominant resource. In these cases it is obvious that the critical resource will be the focus of the technological development process. Thus, the dependence is more or less of a one-way type and the firm has very few possibilities of influencing the resource in any way. In relation to the dependent resource interfaces, learning for a focal firm, in this case Scania, will be focused on how to combine efficiency in providing the dependent resource for the actor possessing the dominating resource, and flexibility and adaptations stemming from demands from actors that control the critical or dominating resources. In essence, research and development needs to take into account the separate alternative scenarios applicable to the dominating resource. This must be maintained to preserve a certain flexibility if one is to respond quickly to changing demands[4]. The owner of the dominating resource will have to gain enough knowledge to understand the consequences of change in a resource related change, or they will have to rely on their counterpart's capability and willingness to adapt to change.

Independent resource interfaces exist when the resource items are two well-known and accepted standards with "open" interfaces. Usually the actors in the industry have developed standardised methods of production over time. In all probability, the standardisation is the result of actors perceiving standardisation as cost efficient. A standardised resource combination or constellation implies that there are more or less clear cut interfaces between the different resources. The interfaces have to be given because of how the resources are purchased and sold. The knowledge required for an actor acting in such a market can be said to be limited to how to produce the resource as efficiently as possible. In the short run, the qualitative knowledge aspect is of less concern, since the actor cannot change the resource's properties as it must fit in to the system in which it is used. Given the independent interfaces, exploring and learning more about the resource network and how it works is, more or less, generally speaking, a waste of time and money. The exception being those relationships that

---

[4]Note that the actor controlling the dependent resource does not have to be a customer. In many cases the focal firm probably finds itself being in the third or even fourth tier in a supplier chain. The focal firm is not required to concentrate on more than the dependent resource in their daily work and its strategic R&D, which of course greatly facilitates a standardized perception of the whole network. The firm controlling the dominating resource is most likely also the actor conducting most of the coordinating work in the network. Therefore, a firm supplying a dependent resource is able to take the development and knowledge generation for granted, as the possessor of the dominating resource takes care of most of the development processes anyway.

provide the learning opportunities for the producer. Contacts with counterparts to develop resources are rarely needed.

Interdependent resource interfaces is the most complex arrangement of resources. When resources are considered to be interdependent, they have been designed specifically to work together within a technological system, and thus if either of the resources were to be replaced, there would be no readily available replacement or substitute that would enable the system to function efficiently. Since the resources are interdependent, they have either been developed simultaneously, or they may have been developed over time through a mutually adaptive, extended development process. When resources are interdependent, there has usually been substantial investment in a constellation of resources. Within the resources, and in particular in the interfaces between the resources, a considerable amount of knowledge has been embedded.

The resources are not only related to each other in a mode of interdependence, but in many instances they are simultaneously interdependent with other resources, thus combining a number of actors in an intricate web: what we commonly call networks of resources. Thus, changing one resource not only affects those directly related resources, but also indirectly affects resources several tiers away. This, of course, means that changes can not be implemented without careful consideration of all resources dependent on the focal resource. This does not necessarily imply that resources will not change as frequently as in other networks. Instead, changes are continuously being made incrementally, and resources are constantly adapted to each other. It does, however, imply that knowledge needs to be dispersed systematically throughout the network for implementation to be successful.

In taking care of the heterogeneity of the resources, relationships between two actors are the most effective way to handle interdependent resource ties. By getting involved in relationships, an actor can influence the way a resource develops. By staying out of them, the actor will meet the future as 'a random walk'. It is, however, not just a matter of immersing oneself in a relationship; an actor has to control some resources of interest and/or knowledge to gain the attention of other actors.

Conclusion

The case illustrates a gradual shift from a dependent resource interface to an interdependent resource interface. This change in technological conditions, spawned by the introduction of computerised systems into truck engines, virtually forced Scania to develop a new partnership if they wanted to continue spearheading the development of truck engines. Together with Cummins they found the opportunity to work in a relationship where the knowledge input of both partners was needed in order to develop new technology. The long-term benefit in engaging in this new relationship was evident for both partners. This is further augmented by the fact that Bosch started to develop new fuel injection systems, albeit with other actors in the market. Too often companies are seen as entities that can choose both the technology and the counterparts freely. Of course, we believe that companies are able to make decisions. However, there is no total "freedom of action". Companies act within the framework within which the picture is painted. This means that, to some extent, solutions are pre-determined, but within the framework provided. That is, within the limitations posed by environmental constraints, such as technology, actors can consciously choose to act freely. Firms act within a network logic, that both de-limits and increases the firm's strategic opportunity space. This frame is not given, but can in turn be affected by an acting company. One important consequence of this is the ability to read the frame and to be an active actor within it.

References

ADERMALM, L, SJÖBERG, H., and WEDIN, T., 1998, Factors behind outsourcing. Illustrations from Scania and Hifab International. MBA Thesis, Department of Business Studies, Uppsala university.

ALCHIAN, A. A., and DEMSETZ, H., 1972, Production, Information Costs and Economic Organisation *The American Economic Review*, Vol. 62, pp. 777–795.

ANDERSON, J., HÅKANSSON, H., and JOHANSON, J., 1994, Dyadic Business Relationships Within a Business Network Context *Journal of Marketing*, Vol. 58, 1994, pp. 1–15.

DEMSETZ, H., 1988, The Theory of the Firm Revisited *Journal of Law, Economics and Organization*, Vol. 4, No. 1, pp. 141–161.

HÅKANSSON, H., 1989, *Corporate Technological Behaviour – Cooperation and Networks*, London: Routledge.

HÅKANSSON, H., and SNEHOTA, I., 1995, *Developing Relationships in Business Networks*, London: Routledge.

HIPPEL, E. VON, 1988, *The Sources of Innovation*, New York: Oxford University Press.

LAAGE-HELLMAN, J., 1997, *Business Networks in Japan; Supplier –Customer Interaction in Product Development*, London: Routledge.

PENROSE, E. T., 1959, *The Theory of the Growth of the Firm*, Oxford: Basil Blackwell.

UTTERBACK. J. M., and ABERNATHY, J., 1975, A Dynamic Model of Process and Product Innovation *OMEGA*, Vol. 3, No. 6, pp. 639–656.

WYNSTRA, F., 1997, *Purchasing and the Role of Suppliers in Product Development*. Licentiate Thesis, Department of Business Studies, Uppsala university.

CHAPTER 8

The Usefulness of Network Relationship Experience in the Internationalization of the Firm

ANDERS BLOMSTERMO, KENT ERIKSSON, JAN JOHANSON, and D. DEO SHARMA

Introduction

In the literature on the internationalization process, it is emphasized that market knowledge is important (Bilkey and Tesar 1977; Cavusgil 1980; Erramilli and Rao 1993; Luostarinen 1980; Makino and Delios 1996). It has also been demonstrated repeatedly that relevant market knowledge is gained through the experiences obtained through doing business in the markets (Johanson and Vahlne 1977; Erramilli 1991; Yu 1990). It is, however, not clear what kinds of market experience are useful in internationalization. Previous research shows that firms primarily act within their existing business relationships, and, when trying to expand their markets, they frequently base their expansion on these relationships (Sharma and Johanson 1987; Erramilli 1991; Hellman 1996; Chen and Chen 1998). Studies also suggest that experience from various kinds of market network relationships is useful when entering into and expanding in international markets (Axelsson and Johanson 1992; Blankenburg 1995; Coviello and Munro 1997; Chen and Chen 1998). Such business relationships may involve domestic and foreign suppliers and customers, as well as the surrounding network of firms, such as customers' customers, and competing suppliers. International expansion can thus be studied as inter-action in international business relationships.

Anders Blomstermo, Kent Eriksson, Jan Johanson and D. Deo Sharma

A number of studies have demonstrated that the experience of doing business in one relationship may be useful in the development of other relationships (Håkansson 1982; Sharma and Johanson 1987; Johanson and Mattsson 1988; Håkansson and Snehota 1995; Blankenburg-Holm, Eriksson, Johanson 1996). It seems that relationships can best be understood in the context of connected business network relationships (Blankenburg 1995). In fact this observation is one of the cornerstones of the network approach to business market management. Although the observation is well documented, it has only been reported in anecdotes and case studies (Håkansson and Snehota 1995; Lee 1991; Hertz 1993).

Learning is another important theme in discussions concerning the internationalization of the firm. It is often argued that internationalization is influenced by the duration of international experience as well as by the variation there has been in this experience (Erramilli 1991; Barkema and Vermeulen 1998). A number of studies even use duration of international business operations as an indicator of experience (Bilkey and Tesar 1977; Czinkota 1982; Davidson 1983; Terpstra and Yu 1988; Root 1994). The stress on duration and variation is consistent with organizational learning theory. Surprisingly, no attention has been paid to how the firm develops its ability to use its network over time. Against this background we examine how the usefulness of network relationship experiences is influenced by the duration of and variation in international operations.

The purpose of this chapter is to explore the role of network relationship experiences, and then to see how such experiences are affected by duration and variation. More precisely, the objective of the exploratory analysis is to find out whether it is possible to form valid constructs based on variables indicating the usefulness of the experience gained from relationships with specific categories of connected firms. It can be assumed that experiences are useful to the extent that they have an effect on firms' actions whilst they are internationalizing. Thus the chapter emphasizes the behavioral aspects of learning whilst undergoing internationalization.

After describing the database in the first section of the paper, we go on to present and comment on the respondents' answers to questions about the usefulness of network relationship experiences for their firms' specific commitments to the development of business with one or more firms in a foreign market. In the subsequent section, we explore the data further by developing and investigating a LISREL model of the usefulness of network relationship experiences. For this purpose we also posit that the usefulness of various network relationship experiences is affected by the duration and variation of the firms' prior international business operations. The analysis indicates that two different constructs capturing the usefulness of network relationship experience are valid. In a further step we also outline two structural models of the relation between duration and variation as independent constructs and the two experience constructs as dependent variables. On the basis of these models we discuss some of the implications for research and management.

The empirical material to be explored

Data were gathered by questionnaire as part of the ongoing research project *Learning in the Internationalization Process*. A pilot study was conducted in 1997 in which ten Swedish firms with international operations were asked to answer a questionnaire in an interview situation. The final standardized questionnaire was sent out in 1998 to managing directors in Swedish manufacturing and service industries with international operations. The Swedish Trade Directory was used to find the addresses of suitable companies. Consequently, the sample is not an independent random one. The firms vary in size, industry sector and geographical location. 176 questionnaires were returned. The response rate was approximately 35%. All the questions were of a closed-ended nature, using a seven point Likert scale ranging from completely agree to completely disagree. Most of the variables are perceptual measures, but there are also objective ones.

In this paper the focus is upon a specific international business assignment. Respondents were asked to a business assignment that was important to their firm and through which their company was expanding internationally. This assignment should preferably be well underway so that the company would already have started doing business with the counterparts. If this was not possible, the respondent was asked to choose a recently finished assignment. Examples of such assignments are:

- A contract with a new distributor or agent in a new country;
- A considerable expansion of the business conducted with an existing customer;
- Doing business with one or more new customers within an existing market;
- Entering new markets abroad with existing customers;
- Doing business with new customers within a new market.

This chapter focuses on a set of statements about the usefulness of the business experience gained from doing business with a number of different customers and suppliers with which the respondent should agree or disagree. The wording of the statements is: In developing this particular assignment, it is useful to have had previous business experience with:

Anders Blomstermo, Kent Eriksson, Jan Johanson and D. Deo Sharma

Questions	ABBREVIATION[1]
Customers[2] in Sweden	CIS
Customers abroad	CA
Suppliers in Sweden	SIS
Suppliers abroad	SA
Customer's customers	CC
Customer's suppliers of products and services that supplement yours	CSS
Competing suppliers	CS

[1]Abbreviation is used in the LISREL analysis
[2]No differentiation is made between customers, distributors and agents.

Tentative exploration

The respondents' evaluations of the usefulness of experience of the seven different types of network relationship are given in Table 8.1, together with the mean values and variance. Looking at the means of the network relationships, we find that having experiences of customers abroad and having experiences of domestic customers are considered to be most useful, with mean values of 2.38 and 3.25. Not surprisingly, the experiences of working with customers are considered to be very useful, implying that the experiences can be transferred from a relationship with one customer to relationships with others. Customer experiences may consist of knowledge of what kinds of products and services customers consider important in different situations. As buyers become demanding, the supplier is forced to learn about customer-specific needs. This seems to be consistent with the notion that customer following is a viable strategy when expanding internationally (Majkgård and Sharma 1998). However, the variance in the experience of domestic and foreign customers does differ. Customers abroad are always useful, clarifying why the variance is very low. Experiences with domestic customers cannot always be applied, which is why very few respondents mark the middle of the seven-point scale. The fact that few respondents score in the mid-range gives us reason to believe that there are two distinctly different kinds of assignments with regard to the usefulness of domestic customer experience.

The table shows that in terms of the usefulness of supplier experiences, there is little difference between the experience of conducting business with domestic and foreign suppliers. The means are 4.07 and 4.13, but these particular averages are a statistical artefact: the respondents have either indicated that they agree or they disagree, but very seldom do they give an answer that lies in between. Thus, we have reason to conclude that the use of supplier experiences in foreign assignments is of two distinctly different kinds. One uses supplier experiences, and one does not.

Experience of the customers' customers and supplementary suppliers are of roughly equal usefulness: 4.25 and 4.43, respectively. However, their

TABLE 8.1
Perceived usefulness of different kinds of network relationship experiences

Question	1	2	3	4	5	6	7	Tot	Mean	Var
CIS	27.4	20.6	14.3	9.1	6.3	10.9	11.4	100	3.25	4.42
CA	33.1	36.0	13.1	7.4	2.3	4.0	4.0	100	2.38	2.48
SIS	14.3	13.1	12.0	14.3	9.1	17.7	18.9	100	4.13	5.26
SA	14.9	14.9	13.1	15.4	8.6	13.1	20.0	100	4.07	4.50
CC	13.2	12.6	9.2	20.7	7.5	19.0	17.8	100	4.25	4.19
CSS	4.6	12.6	17.2	21.8	7.5	16.7	19.5	100	4.43	3.43
CS	9.7	22.9	18.9	17.7	10.3	7.4	13.1	100	3.71	3.71

Note: Frequencies are in valid percent. 1 is fully agree and 7 fully disagree.
CIS = The usefulness of previous business experience with customers in Sweden
CA = The usefulness of previous business experience with customers abroad
SIS = The usefulness of previous business experience with suppliers in Sweden
SA = The usefulness of previous business experience with suppliers abroad
CC = The usefulness of previous business experience with customers' customers
CSS = The usefulness of previous business experience with customers' suppliers of products and services that supplement yours
CS = The usefulness of previous experience with competing suppliers

variance does differ. Experience gained through conducting business with customers' customers is either useful or not – seldom in between – while supplementary suppliers are seldom useful.

According to the means, the next most useful experiences after that of doing business with customers are gained from competitors. Such experiences are almost always useful, although not to the same extent as customer experiences.

The overall impression given by Table 8.1 is that all the different network relationships provide useful experiences sometimes, and that most of them prove useful quite often. It appears that experiences gained from different kinds of network relationships have similarities as well as differences. The fact that there is mixed usage of the different network relationships probably reflects that firms are in specific situations. To clarify this, we take a further step in analyzing the usefulness of the different sources of experiences.

The technique for exploration using LISREL

In the next step we examined whether it was possible to form a smaller set of meaningful constructs that capture much of the variation in the variables described in the previous section. This was done with the help of the LISREL method, a technique for tracing structural relations in a data set (Jöreskog and Sörbom 1993). The technique is unique in that it puts stringent requirements on validity by using the correlation estimate and the

Anders Blomstermo, Kent Eriksson, Jan Johanson and D. Deo Sharma

correlated error terms as two independent sources against which patterns of variation in the data are tested. Even though these stringent requirements are sometimes difficult to meet, they enable advanced tests to be made of causalities in a data set. The technique also tests for constructs, which are higher-order representations of the common, underlying commonalities that have been observed in a set of indicators.

The validity assessment is done to ensure that the constructs are indeed independent from one another, and to decide whether, when taken together they make up a valid pattern of structural relations in a model. The LISREL program gives the user powerful tools for assessing this validity. Perhaps most helpful is the modification index, which presents a graphical interpretation of the strongest disturbances in the model. It may, for instance, be revealed that two constructs load on the same indicator, which means that one cannot discriminate between them. In other cases, patterns of correlations between error terms may reveal that the relation between two indicators needs to be under-represented in the model for it to be valid.

Exploration of patterns of variation in the data with LISREL hinges on the use of the graphical modification index, which helps the researcher to re-evaluate or arrive at theoretical explanations. The common procedure is for the researcher to hypothesize a model, and then change it in the light of modifications suggested and substantive theory. At best, the dialectic between the empirical material and theory provides fertile soil for a better understanding of empirical material and more developed theory. However, there is a need to recognize that the final results should rest on firm theoretical grounds if they are to be acceptable. Valid patterns in data are not enough in themselves.

The exploration using LISREL

It takes time to develop unique relationships, and thereby different kinds of market knowledge. This kind of knowledge is a product of the duration the firms have been involved in international business operations, and the variation of their business operations. Variation in international contexts is necessary for the learning capacity of the firm. According to the internationalization process, variation is an intermediate construct between duration and experiential learning, i.e., variation facilitates foreign experiential knowledge development, but it takes time to build operations in different countries and, consequently, to create variation.

Duration is often measured as an objective question concerning the time for the first international operation. In line with that, we measure duration by asking managers: which year did the firm start doing international business?

Variation is often measured as the geographical dispersion of the firm (Erramilli 1991). Consequently, variation was measured by asking the respondents: to how many countries do you sell?

Table 8.1 shows that some types of business relationships are more useful than others. The explorations start out from the assumption that firms working with an assignment probably use the experiences gained from previous business relationships. However, the tentative results shown in Table 8.1 are not taken as the starting point because LISREL provides an opportunity to analyze data with a second, independent, and more advanced technique. It is therefore hypothesized that all seven indicators of customer relationships form one construct that correlates with duration and variation. Such a model is depicted in Figure 8.1 below.

FIGURE 8.1
Path diagram displaying hypothesized structural LISREL model

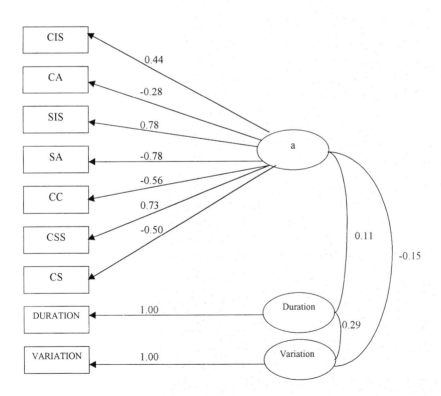

The model taken as the starting point shown in Figure 8.1 is invalid. This suggests that we need to change it with the help of the modification index provided by LISREL. As a starting point for exploration, it is sometimes fruitful to add modifications until the model is valid. This may result in complicated structures that it is almost impossible to interpret, but it may also give clues to further work that it would be advantageous to pursue. The model in Figure 8.1 needed five modifications in order to become valid. It should be noted that these are the five strongest disturbances in the model. In other cases, it may be desirable to modify the model by introducing those modifications that are conceptually relevant. But the conceptual foundation of the model hypothesized here does not require that the researcher selects conceptually relevant modifications at this stage of the exploration.

The five modifications are displayed in a valid model in Figure 8.2 below. Each arrow or line signifies a relation that does not fit with the hypothesized model. As can be seen, most arrows start or end at the first four indicators. Apparently, there is some disturbance between these four variables which makes them unsuitable for use with the others. One possible conclusion could be that the first four indicators, which deal with the firms' suppliers and customers, differ from the last three indicators, which focus customers' customers, complementary suppliers, and competitors. It may be that the first four are more related to the respondent than the last three. However, this is a tentative finding which requires further analysis.

The modifications suggest that a more limited exploration of these four indicators of suppliers and customers is needed to understand them better. The first test is to combine them to form one construct. The model could not even be generated using these four indicators because the tensions between them were too strong. Several attempts to make two constructs, with two indicators in each, also failed. So, the result of explorations of the first four indicators is that they neither lend themselves to the formation of one construct nor two.

To analyze this finding, one needs to explain two complicated concepts behind LISREL modeling. The first is the concept of nested structures. Consider four or more indicators which are related to one another so that A is strongly related to B, B is strongly related to C, and C is strongly related to D, but all other relations between these indicators are weak. These indicators are nested in the sense that there are multiple substructures between them that make it difficult to combine them into one construct, and also difficult to separate them into two or more constructs. A, B, C and D do not form one construct since the relations between the indicators are simultaneously weak and strong. There is no ground for viewing them as one valid construct since they do not share underlying commonalities. However, a model with two constructs consisting of A and

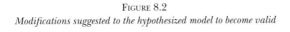

FIGURE 8.2
Modifications suggested to the hypothesized model to become valid

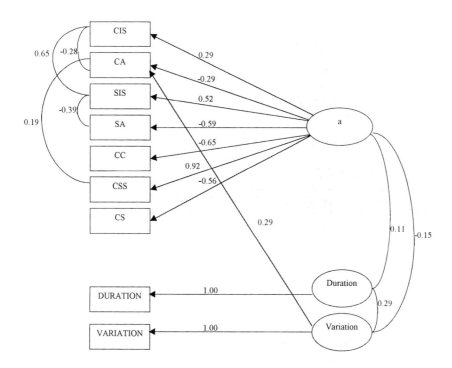

B on the one hand, and C and D on the other, will not be valid either. This is because the strong relationship between B and C demands that the model be modified.

The second complicated issue concerns the more detailed interpretation of modifications. A modification is a disturbance to the model, that is, the model would be acceptable were it not for the disturbances identified in the modifications. Since the model identifies a pattern of variation derived from a data set, a modification can be interpreted as a subpattern of variation in the data that does not quite fit with the model. Almost like an anomaly to the model. Or, put differently, the model fits well with the data, except for the modifications suggested. Such exceptions can occur because the exception is a stronger or weaker deviation in the data. This means that a model that includes a modification gives a slightly distorted view of data, and that this distortion causes some relations in the model to be stronger or weaker than they are in the real data set. If the modification is positive, then it means that the relation between these indicators is stronger in the data than in the model, which implies that the relation is underspecified, or weaker in the model than in the data. Negative modifications are overspecified, or stronger in the model than in the data. Together with the insight that structures can be nested, modifications provide a useful tool for further exploration of the material currently under examination.

The reason for the problems associated with the four indicators on suppliers and customers is probably that the indicators are nested. Figure 8.2 shows negative modifications between domestic and foreign suppliers, and between domestic and foreign customers. This means that the relation in each of these pairs of indicators is stronger in the model than in the data. An additional piece of information gives us sufficient grounds to draw conclusions: the relation between domestic customers and suppliers is weaker in the model than in the data. The nested structure can be depicted as in Figure 8.3, where solid lines are disturbances to the model. The results suggest that the foreign customers and suppliers should be a separate construct, since the domestic customers and suppliers are stronger in the data set than in the model. However, this model was rejected. Instead, LISREL suggested that the four indicators should be put back into one construct. Apparently, the combination of domestic and foreign with suppliers and customers creates a dual structure that contains tensions, making it difficult to view it as one or several constructs. A likely conclusion is that domestic and foreign are different dimensions from customer and supplier.

Another conclusion possible from this exploration is that there is a nested structure, which is greatly disturbed as it tries to represent the relations between domestic and foreign customers and suppliers. The other indicators for business networks seem to be less nested.

To continue the analysis, one can concentrate instead upon the indicator with the most modifications. This is the CA indicator, which gives information on the usefulness of previous business relationships with customers abroad. Since the design of the questionnaire focuses on customer assignments in terms of conception, this is a central indicator. It can be seen that both the variation construct and supplementary customers (CSS), have relations to customers abroad which are weaker in the model than in the data. The relation between customers abroad and Swedish customers is, on the other hand, stronger in the model than in the data. Thus it seems as if the role of customers based abroad is downplayed by the Swedish customers, and that the model does not incorporate the importance of customers abroad.

Much information has been extracted from the modified model in Figure 8.2. Summarizing this information is instructive as it facilitates the

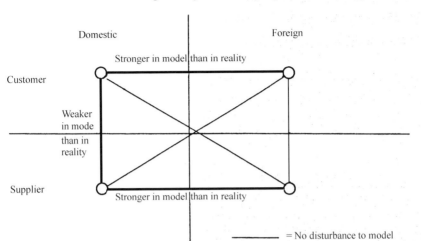

FIGURE 8.3

A nested structure demonstrating the use of domestic and foreign customers and suppliers

drawing of conclusions about what step to take next. The findings so far are that:

- The domestic and foreign customers and suppliers disturb the model since they form a nested structure.
- Domestic customers and suppliers disturb the model strongly.
- The role of customers abroad is downplayed in the model.

One possible conclusion to be drawn from this is that domestic customers and suppliers can be deleted from the model, and that customers abroad can be a construct of its own. As mentioned earlier, various combinations where domestic customers and suppliers are included have not been successful. Apparently, they disturb the model somehow. This result is very interesting and should certainly be the object of further study. The domestic and foreign settings are key to understanding internationalization since firms develop capabilities at home and abroad, and use them at home and abroad. But such studies require an in-depth study of the nested structure, and this is somewhat tangential to the purpose of this chapter. One way to continue is to exclude some of the indicators that cause disturbances. Perhaps they can be included again, once the relations between the other indicators have been better understood.

Trials of the new structure have been successful after the deletion of the indicator that concerns customers' suppliers of products that supplement the responding firm's (CSS). The model can be constructed with the CSS indicator, but the program suggests a modification that goes from the

Anders Blomstermo, Kent Eriksson, Jan Johanson and D. Deo Sharma

indicator to customers abroad (CA). Since their relation is weaker in the model than in the data, it may appear to be reasonable to move the supplementary supplier (CSS) indicator to combine it with customers abroad (CA). However, the resulting model is invalid. Supplementary suppliers can apparently play a role combined with both customers abroad and the other indicators on relationships. Because it is desirable to be able to discriminate between constructs, the supplementary supplier indicator is deleted from the model altogether. The resulting combination of constructs makes up a valid model as depicted in Figure 8.4.

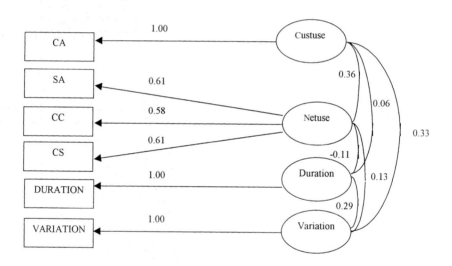

The first construct is the usefulness of foreign customer experiences Custuse in the ongoing assignment. It shows almost no correlation with duration (0.06), and as the relation is actually not significant, it cannot be assessed. Foreign customer experience correlates with both variation (0.33), and the construct foreign network experience, Netuse (0.36).

The construct foreign network experience captures the usefulness of previous experiences with suppliers abroad, customers' customers and competing suppliers. The construct has a significant correlation only with foreign customer experience (0.36), not with variation (0.13) or duration

(−0.11). The correlation between variation and duration is 0.29, and is significant.

The results in Figure 8.4 can be put together as shown in Table 8.2, which displays each construct and its corresponding indicators. Foreign network experience is the only construct that captures all three indicators, the rest of the constructs capture just one single item.

TABLE 8.2

The constructs and their indicators

Constructs and Indicators[1]	Indicator Label	Factor Loading	T-value	R2-value
Foreign customer experience When carrying this assignment out it is useful to have had previous experience of customers abroad	CA	1.00		

Foreign network experience
 When carrying this assignment out it is useful to have had previous experience of
     suppliers abroad
     customer's customers
     competing suppliers
	
	SA
	CC
	CS
		
		0.58
		0.55
		0.66
			
			6.06
			5.83
			6.55
				
				0.34
				0.30
				0.44

Duration Number of years since first international business assignment (log transform)	DURATION	1.00		
Variation of experience Approximately how many countries do you operate in? (log transform)	VARIATION	1.00		
[1]The indicators are identical with the questions put to respondents				

Exploration of structural models containing the constructs

As discussed earlier, it is anticipated that internationalization experience comes from variation in markets, which is an antecedent to the usefulness of previous business relationship experiences. It is also posited that duration determines variation, since internationalization takes time. A structural model was therefore made with duration as an independent variable, effecting variation, which then affects both foreign customer experience and foreign network experience. Figure 8.5 displays such a model. However, the model is not valid, so there is a need to modify it.

The modification indices suggest that the model is valid if there is a causal relationship from foreign network experience to foreign customer experience. Such a model is displayed in Figure 8.6 below. The results in Figure 8.6 clearly show that variation has more effect than duration on the usefulness of relationship experience. The results also show that duration has a strong effect on variation (0.30 in Figure 8.6). This shows that the

FIGURE 8.5
The structural model

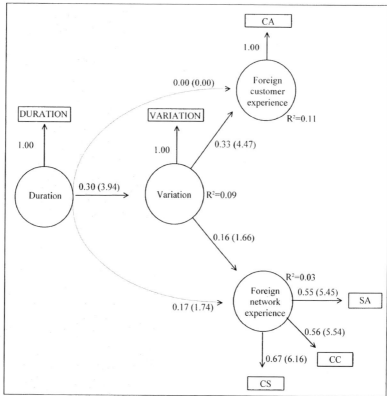

Note: Figures are factor loadings, with t-values in parentheses. Chi-square is 17.74, with 7 degrees of freedom, at a probability of 0.01 (not significant at the 5%-level).

construct duration influences firms' ability to use their experiences, and that variation is necessary for this development to happen. Variation is thus more essential than duration for experience to develop, but variation results from duration.

The results in Figure 8.6 also indicate that variation increases the usefulness of both foreign customer experience (0.28) and foreign network experience (0.22). This shows that variation in the markets in which a firm operates increases a firm's capability of using its prior business relationship experiences. Apparently, firms develop routines for using prior experiences as they gain in international experience.

The results also show that foreign network experience increases foreign customer experience (0.31). This shows that foreign customer experiences are often associated with foreign network experiences. Presumably, the network experiences are a context to a specific relationship, and in many

FIGURE 8.6
The final structural model

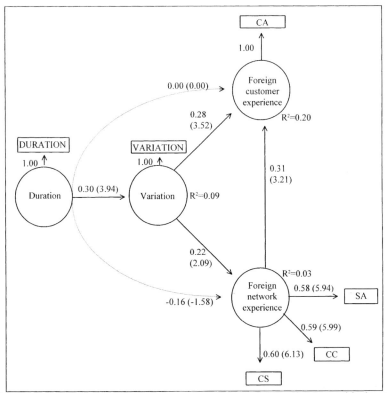

Note: Figures are factor loadings, with t-values in parentheses. Chi-square is 5.50, with 6 degrees of freedom, at a probability of 0.48 (significant at the 5%-level).

cases it is necessary to understand how to use these contextual experiences before the relationship experiences can be used. For instance, it may be that the network experiences are useful in a foreign business assignment that a firm has committed itself extensively to. Or, the other way around, perhaps the early stages of relationship development include a more limited understanding of the context, which then makes the firm less able to use its network experiences.

However, network and relationship experiences are distinct, as shown by the discriminant analysis. It is often argued that the customer relationship and the network are not two distinct analytical dimensions, but that they merge into one complex whole. The present results do not contradict such an argument. Rather, the results of this study are that the past experiences of customer relationships and the networks are used differently. It may well be that the relationship and the network are inseparable entities in ongoing business, but this is not studied here.

The results presented in Figure 8.6 show good statistical key figures in all cases apart from two, these being two poor R2 values that concern the causal relations with variation (R2=0.09) and relationship network experiences (R2=0.03). This means that the linearity of the relationship is poor, but it should be noted that the relationship is significant and reasonably strong.

The implications of the results are that it is difficult to understand how variation and relationship network experiences are caused by their respective independent constructs. This is not unexpected. Frequency distributions in Table 8.1 show that the network relationship variables are not normally distributed. There seem to be firms that consider network relationships useful, and those that do not. The low R2 value and the frequency distributions together suggest that the sample contains two groups of firms. We may speculate that this duality is attributable to the depth of involvement in the ongoing assignment. The questionnaire asks respondents to select an assignment, and then answer questions on the usefulness of customers and the usefulness of network connection categories to this assignment. Presumably, uses differ between an assignment where respondents have a considerable degree of involvement and one where the involvement is low.

Suggestions for further studies

The results above show that firms use their past experiences more as they gain experience from more international markets. This shows that experiences can be applied in different markets. Other results also show that firms with a high degree of internationalization knowledge develop more differentiated knowledge structures in specific markets (Eriksson, Johanson, Majkgård and Sharma 1997). This suggests that experienced firms develop routines for going international, which then increases the rate at which they gain a deep knowledge of specific local markets.

However, this does not automatically imply that experiences can be transferred. There are additional pieces of information that need to be gathered before more firm conclusions can be drawn. An important issue may be the profundity of the experience in the ongoing assignment. It can be expected that firms that are deeply involved in an international assignment become more aware of the ways in which it is unique, and this may cause them to be more discerning about which experiences they can use.

The fact that firms develop knowledge both of many diverse country markets, and from the depth of involvement can be fruitfully combined into a research framework for future studies. It may seem plausible that firms with diverse experiences have learned more about what to look for when developing a specific assignment. Put differently, firms with experience of different countries' markets have knowledge structures relevant to the development of specific assignments that are different from those of firms with experiences from fewer, and homogenous markets. Likewise, it may be expected that the more deeply involved a firm is in an assignment, the more unique and discerning is the applicability of the knowledge they generate from diverse markets. Firms with more diverse experiences may capitalize on their knowledge in the initial development of business assignments. But as firms become more deeply involved in a relationship, the more unique it becomes, and the less applicable the existing knowledge. However, all this is speculation and should be studied more carefully in the future.

A very interesting result from the exploratory analysis is the nested structure, where the firm's relationships with domestic and foreign suppliers and customers was found to be complicated. Although the present study could not resolve this issue, it does indicate a potential topic for future research. The domestic and the foreign settings are key to understanding where firms develop their capabilities, and then we need to understand how they apply these strengths in ongoing business assignments. One possible research subject would be to see how the degree of nesting differs with experience in many countries, on the one hand, and with experience of the ongoing assignment, on the other. A framework could be developed with the depth of involvement in an assignment on one axis, and the experience from various markets on the other. Perhaps the usefulness of relationship and network connection categories is different in each of the four boxes.

Our findings on nested structures may have a few research implications. In our opinion, nested structures may be a root cause of heterogeneity among firms. Nested structures could underlie what has been called 'the architectural knowledge' (Henderson and Clark 1990). Whereas the seven individual kinds of knowledge identified in this chapter may represent 'component knowledge', nested structures represent organization schemes for coordinating the various individual components of knowledge in firms and putting them into productive use. In such a case, the nested structures identified in this chapter make firms unique by making imitation difficult, if not impossible. Whereas any of the seven individual components of knowledge identified in this paper can be imitated by other firms, or can be transferred to other firms, it is difficult to imitate and transfer architectural knowledge binding component knowledge. If this is the case, the nested structures in firms may be the ultimate source of the competitive advantage that they possess. On all these issues more research is greatly needed.

Anders Blomstermo, Kent Eriksson, Jan Johanson and D. Deo Sharma

References

AXELSSON, B., and JOHANSON, J., 1992, Foreign market entry – the textbook *vs.* the network view, in Axelsson, Björn, and Easton, Geoffrey, (eds), *International Networks: A New View of Reality*, pp. 218–234 London: Routledge.

BARKEMA, H. G., and VERMEULEN, F., 1998, International expansion through start-up or acquisition: A learning perspective *Academy of Management Journal*, 41, 1: pp. 7–26.

BILKEY, W. J., and TESAR, G., 1977, The export behaviour of smaller sized Wisconsin manufacturing firms *Journal of International Business Studies*, 8(1): pp. 93–98.

BLANKENBURG, D., 1995, A network approach to foreign market entry, in Möller, K., and Wilson, D., (eds) *Business Marketing: An Interaction and Network Perspective*, pp. 375–405, Boston: Kluwer.

BLANKENBURG-HOLM, D., ERIKSSON, K., and JOHANSON, J., 1996, Business networks and cooperation in international business relationships, *Journal of International Business Studies*, 27(5): pp. 1033–1053.

CAVUSGIL, S. T., 1980, On the internationalization process of firms, *European Research*, 8(November): pp. 273–81.

CHEN, H., and CHEN, T.-J., 1998, Network linkage and location choice in foreign direct investment *Journal of International Business Studies*, 29(3): pp. 445–468.

COVIELLO, N., and MUNRO, H., 1997. Network relationships and the internationalisation process of small software firms *International Business Review*, 6(4): pp. 361–386.

CZINKOTA, M., 1982, *Export development strategies: US promotion policies*, New York: Praeger Publishers.

DAVIDSON, W. H., 1983. Market similarity and market selection: Implications of international marketing strategy *Journal of Business Research*, 11: pp. 439–56.

ERIKSSON, K., JOHANSON, J., MAJKGÅRD, A., and SHARMA, D., 1997, Experiential knowledge and cost in the internationalization process *Journal of International Business Studies*, 28(2): pp. 337–360.

ERRAMILLI, M. K., 1991, The experience factor in foreign market entry behavior of service firms *Journal of International Business Studies*, 22(3): pp. 479–501.

_____ and C. P. Rao, 1993, Service firms international entry mode choice: A modified transaction-cost analysis approach *Journal of Marketing*, 57: pp. 19–38.

HÅKANSSON, H., and SNEHOTA, I., 1995, *Developing Relationships in Business Networks*, London: Routledge.

HÅKANSSON, H. (ed.), 1982, *International Marketing and Purchasing of Industrial Goods: An Interaction Approach*, Chichester: Wiley.

HELLMAN, PASI, 1996, The internationalization of Finnish financial service companies *International Business Review*, 5(2): pp. 191–208.

HENDERSON, R. M., and CLARK, K. B., 1990, Architectural innovation: The reconfiguration of existing product technologies and the failure of established firms *Administrative Science Quarterly*, 35(1): pp. 9–22.

HERTZ, S., 1993, The Internationalization Process of Freight Transport Companies – Towards a Dynamic Network Model of Internationalization, Stockholm, EFI at the Stockholm School of Economics (Dissertation).

JOHANSON, J., and VAHLNE, J.-E., 1977, The internationalization process of the firm – A model of knowledge development and increasing foreign market commitments *Journal of International Business Studies*, 8(1): pp. 23–32.

JOHANSON, J., and MATTSSON, L.-G., 1988, Internationalization in industrial systems. A network approach, in Hood, N., and J.-E. Vahlne, (eds) *Strategies in Global Competition*, pp. 287–314, New York: Croom Helm.

JÖRESKOG, K.-G., and SÖRBOM, D., 1993, *LISREL 8: Structural Equation Modeling with the SIMPLIS Command Language*, Chicago: Scientific Software Internationals.

LEE, J.-W., 1991, *Swedish Firms Entering the Korean Market – Position Development in Distant Industrial Networks*, Uppsala: Department of Business Studies.

LUOSTARINEN, R., 1980, *Internationalization of the Firm*, Helsinki School of Economics: Helsinki.

MAJKGÅRD, A., and SHARMA, D. D., 1998, Client-Following and Market-Seeking Strategies in the Internationalization of Service Firms, *Journal of Business-to-Business Marketing*, 4(3): pp. 1–41.

MAKINO, S., and DELIOS, A., 1996, Local knowledge transfer and performance: Implications for alliance formation in Asia *Journal of International Business Studies*, 27(5): pp. 905–927.

ROOT, F. J., 1994, *Foreign Market Entry Strategies*, New York: AMACOM.

SHARMA, D. D., and JOHANSON, J., 1987, Technical consultancy in internationalisation, *International Marketing Review*, 4(Winter): pp. 20–29.

TERPSTRA, V., and YU, C-M., 1988, Determinants of foreign investment of U.S. advertising agencies, *Journal of International Business Studies*, 19(1): pp. 33–46.

YU, C-M. J., 1990, The experience effect and foreign direct investment, *Weltwirtschatliches Archiv*, 126: pp. 561–79.

CHAPTER 9

Expectation – The Missing Link in the Internationalization Process Model

AMJAD HADJIKHANI and MARTIN JOHANSON

Introduction

The increase in the number of international business studies concerning market exit and contraction are evidence of a growing rate of market turbulence (Makhija, 1993; Nilson, 1995). However, some researchers have been critical of the prevailing conceptual tools and have questioned their relevance (Fink, 1986; Kauzman and Jarman, 1992). They argue that turbulence, that is, unpredictable changes in the environment, creates a need for new concepts. Evidently such turbulence is a great problem for firms operating in international markets. In this spirit this paper examines how the internationalization process model (Johanson and Vahlne, 1977, 1990) can accommodate turbulence in the market environment of internationalizing firms. The model has been revised and criticized in a number of studies. There have also been efforts to explore the basic variables of the model – knowledge and commitment – in order to analyze business behavior in turbulent markets (Hadjikhani and Johanson, 1996). Following this track, the aim of this study is to add knowledge and enrich the model by including the variable of expectation.

After a review of the internationalization process model (IP-model) and some studies carried out in accordance with its tenets, we highlight imbalances between market commitment and market knowledge. We argue that the model has problems in handling turbulent markets, as it implicitly assumes a balance between the firm's market commitment and market knowledge, which is seldom the case in turbulent markets. The IP-model considers past and present dimensions of firms' internationalization, but misses the future dimension. The foundation of this paper is the assumption that experiential knowledge is not sufficient to explain the behavior of firms in turbulent foreign markets. The expectation concept embellishes the model with a future dimension. We then present two longitudinal case studies of two Swedish firms' behavior in two different, but changing and turbulent markets – Iran and Russia. The IP-model views internationalization as an incremental process and it is therefore appropriate that the studies employ a process perspective. We conclude the paper by incorporating the expectation concept into the IP-model and discuss the relationship between the variables of expectation experiential knowledge, and commitment decisions in the two case studies.

A review of the IP-model

The base for the IP-model was Johanson and Wiedersheim-Paul's study (1975), where they argued that Swedish firms had internationalized systematically. In order to manage perceived market uncertainty the firms followed a sequential process (no regular export activities, export via independent representatives, sales subsidiary, and production/manufacturing). The market uncertainty was thought to be a direct result of the cultural differences between countries. This study was followed by the IP-model (Johanson and Vahlne, 1977). The model had two groups of variables – the state and the change aspects – that were divided into market commitment and market knowledge and current business activities and commitment decisions respectively. The essence of the model is that the point of departure for an extended degree of internationalization is partly the knowledge the firm has about the specific foreign market and partly the resources that are committed to that market. Another important aspect is that market knowledge is of two types. One is experience-based and can only be developed by running business activities in the foreign market, whereas the second type, objective knowledge, can be taught and transferred between individuals and firms. Moreover, current business activities and commitment decisions positively affect market commitment and market knowledge. Current business activities are the main source of experiential knowledge. Commitment decisions have two effects. The economic effect is an increase of the scale of operation, whereas the uncertainty effect concerns uncertainty that the firms perceive. Johanson and Vahlne define market uncertainty as the perceived inability to estimate the present and future market and market-influencing factors. Uncertainty is reduced through interaction and integration with the foreign market environment.

Amjad Hadjikhani and Martin Johanson

A comment on the IP-model

There are two main reasons why the IP-model has problems in handling firms' operations in turbulent markets. *First*, it is a model of increased market commitment and knowledge development, that is, a model for an extension of the firm's operation in markets other than the domestic one. The model does not claim to deal with a decrease of commitment and knowledge, which is currently very much a reality. It should not be criticized in that respect. However, firms operating in foreign markets with a high commitment to some specific market sometimes decide to decrease their commitment, or even withdraw from the market and try to transfer resources to other markets. We believe that negative expectation about the market where the firm is operating can cause it to decrease its resource commitment and even exit the market. Furthermore, the model does not give us any clues about why firms enter or leave foreign markets. Entry is, by definition, an increase in commitment of resources to a specific market, but without any experiential market knowledge. It is likely, however, that the firm would have objective market knowledge. This can be gathered through research and studies, but as important is what managers hear and read from others: i.e. exogenous factor. Since the IP-model emphasizes experiential rather than objective knowledge, and the latter, exogenous, is all that the entering firm can rely on, it has difficulties in explaining its entry into the market.

Second, the extension of the firm's international operations are incremental in nature, sometimes through sequences or step by step, each sequence or step being characterized by a balance between market commitment and market knowledge. However, there are situations characterized by an imbalance between market commitment and market knowledge. For instance, a number of studies have documented that sometimes firms make large market commitments without having any experiential knowledge. It is obvious that firms disregard gaining knowledge incrementally and leap into the unknown. It does not mean that experiential knowledge is unimportant, but that in some cases exogenous factors are the driving forces. In such cases, where the success of other firms may encourage firms to make large investments, the role of expectation becomes more obvious. Expectation connects the future dimension of commitment to decisions made in the present. Indeed, a commitment decision is constructed of the two interwoven parts of past knowledge and future beliefs. In terms of the model, such behavior is naturally tainted by high risk. The important question is why they make commitment decisions with high risk. An appropriate view for the explanation of such behavior is the market expectation. Another example of imbalance, that is also very much a reality for multinational firms, is that their knowledge erodes or becomes obsolete due to market turbulence.

Both institutional and business knowledge can become obsolete and the changes and unpredictability of market conditions can force the firm to make new decisions about whether to keep the commitment intact or to reduce it. The opposite, which is that firms have a high level of market knowledge but are not highly committed to a market, also happens. A firm can have a strong position with weak competition in a small or undeveloped market. A long-term presence in the market has provided substantial experience but has not required any large investments and the firm has consequently decided not to extend its operation. In these cases, when the firm's market knowledge becomes obsolete the decision as to whether to keep the commitment intact or to decrease it is built only on expectation.

A few efforts have been made to deal with these shortcomings. For instance, Axelsson and Johanson (1992), Blankenburg-Holm and Johanson (1992), Johanson and Mattsson (1988), and Johanson and Vahlne (1990) applied a network model to the entry of firms in foreign markets and found that internationalization is a time-consuming process where the entering firms interact with specific firms in the foreign market and gradually develop knowledge and commit resources vis-à-vis those firms. Furthermore, Hadjikhani and Johanson (1996) found that firms use different strategies when they face turbulent markets. Turbulence obviously causes an imbalance between the firm's market commitment and market knowledge. With a large commitment and lack of knowledge firms have to react to changes by relying on more general knowledge and a long-term market expectation. In a study of Swedish firms in a turbulent market undergoing radical economic, political, and social change, namely Iran, Hadjikhani (1996) found that firms' strategies and behavior differed as a consequence of the character of their commitments. When the firms lost their market knowledge due to turbulence, it was evident that only those firms with a tangible commitment, which was more transferable and short-term oriented, preferred to exit the market. On the other hand, firms with a strong intangible commitment continued their operation. Thus, an important implication and comment on the IP-model concerns the imbalance between market commitment and market knowledge caused by turbulent market conditions. The imbalance can either consist of strong commitment and lack of knowledge or weak commitment and comprehensive knowledge.

Case studies

Following the IP-model, the case presentation is longitudinal in nature. The subsequent discussion attempts to clarify factors that increase an understanding of expectation in the internationalization process. The reason for structuring the cases in terms of three time periods is the similarity in the firms' expectations, at least, in the first two time periods. The two historical cases concern two Swedish multinationals, Volvo in Iran and Karlshamns in Russia. The cases illustrate expectation development in three different time periods. In one case, the development of expectation is shown in the periods of increasing expectation, negative expectation development, and finally the exit decision (for the first case) and escalation of market decision (for the second case). In the second case, the process of expectation development was manifested during periods of penetration and expansion. A dormant period followed and finally, a period of reinternationalization.

The empirical data were collected exclusively through in-depth personal interviews. The interviews were based on an open question guide, as the complexity and unpredictability of the subject studied required openness and flexibility during the interview process. In order to be able to keep up with events as they occurred, the interviews were undertaken continuously and there were always two main foci. The first was what the respondents believed, expected, and planned for the future, and the second, what they had experienced since the previous interview. In the following section, the case presentation is structured to show the role of expectation change in the firms' business behavior. The cases consider two different firms, in two different markets, as well as two different time periods. The reason for structuring the cases in terms of three time periods is the similarity in the firms' expectations in each of the periods. In the first period, the market expectation for both firms is increasing. In the second period, expectation develops negatively for both firms. In the third period, their expectations develop interestingly. One firm loses hope and the negative expectation forces it to exit from the market. For the other firm, market expectation gradually gains positive momentum and the firm consolidates its market commitment.

The case of Karlshamns in Russia

Period 1: 1986–1989

Positive expectation is increasing: Before this period, Karlshamns had limited experience of international operations, although they had been exporting cocoa-fat substitutes to the Soviet Union. The yearly sales on average amounted to just 1 or 2% of total turnover. Karlshamns' activities had consisted of maintaining contacts with organizations under the auspices of the Ministry for Foreign Trade. A few salesmen had been responsible for the Soviet market and had all the knowledge about that market. In 1986, Karlshamns was aware that conditions in the Soviet Union were changing, but only had diffuse ideas about what the changes in the market would mean. The strategy for international operations did not include the Soviet

Union, and instead resources were transferred to the United States (USA) and the European Union (EU). During 1997, a number of unusual and unexpected contacts with the Soviet Union arose. Karlshamns' top management decided to keep the door open and the unstructured contacts continued. Karlshamns saw a licensing agreement as a possible approach when in March 1987, they had a meeting with the Minister of Agriculture, Mr Zubkov, and he first mentioned the idea of forming a joint venture. Karlshamns viewed the Soviet Union as the equivalent of a lottery ticket and discussed how much they would be prepared to pay for it. They were fully aware that everything could go wrong, but also that the political transformation process had an end and that the existing system could be replaced by some kind of market economy. They thought the Soviet market had enormous potential in the long term. Karlshamns' aim was to keep the risk low. They determined, first, that in order to minimize the financial risk they should not provide any hard currency as equity for the joint venture and second, that it was crucial that they be able to receive a cash flow in hard currency. Despite misunderstandings and problems during the negotiations, a contract was closed at the beginning of July 1989. The joint venture was called Viking Raps and was planned to be the biggest joint venture in the Soviet Union. The investment was estimated at 2 billion SEK.

Karlshamns did not expect the coming tasks to cause any delays or problems. The Soviet partners were assumed to have all the necessary contacts with the Soviet authorities and consequently, to be able to take care of the registration of Viking Raps at MINFIN. During 1989, Karlshamns had been in contact with Svenska Handelsbanken and SE-banken and both had hinted at an interest in financing the project. They only required a guarantee from Vneshekonombank. Karlshamns' Soviet partners in Viking Raps had said that none of the issues would be a problem and Karlshamns thought that their partners would quickly resolve all these questions. Viking Raps also began to form a project organization. Non-Russians with experience in the Soviet Union and in operations in an international environment were needed. It was decided that Karlshamns should find an established firm with international experience. It took 40 days for MINFIN to accept the application for registration and then it informed Karlshamns that there were 25 unacceptable items in the agreement. In December 1989, the Deputy Minister of Finance visited Sweden and Karlshamns arranged a meeting with him. Six days later Viking Raps was registered. In April 1990 Karlshamns for the first time managed to arrange a meeting with Vneshekonombank. Karlshamns informed them that if they did not get an answer before August 1 they would have to freeze the project. In July 1990 Mr Zubkov and Vneshekonombank agreed that the latter should provide a guarantee if 15% of the stock could be provided in hard currency. Prodintorg, a foreign trade organization, bought 10% of the stock for 55 million SEK and Karlshamns agreed to provide 50 million SEK. On September 9 Vneshekonombank submitted the letter of guarantee.

Period 2: 1989–1991

Negative expectation development: In December 1989, Karlshamns started negotiations with IC, but in February 1990, Karlshamns was told that GIP-3 had the sole right to develop and construct plants for oils and fats in the Soviet Union. Accordingly, that responsibility was given to GIP-3, but it did not have sufficient knowledge or contacts and Karlshamns once more turned to IC. However, in May 1991, IC was again informed that its services were not necessary. The person selected to work with Viking Raps did not "fit in". In the spring of 1990, Viking Raps established contact with two construction firms in Lipetsk – Lipetskstroi and Promstroi – and eventually decided to use Lipetskstroi for the construction work. Due to delays with payments the Swedish banks stopped the granting of credit to projects in the Soviet Union around New Year 1989–90. Karlshamns prepared an application for EKN, which was postponed as it looked like a doubtful venture. From then onwards EKN postponed all applications regarding the Soviet Union. In October 1990 one representative from Karlshamns went to Germany, where he presented the project to one supplier, LURGI, and to a bank, Kreditanstalt. They were positive and intended to apply to Hermes, the German equivalent of EKN. On February 6 1991, Karlshamns received a letter from Hermes stating that it intended to support the project with a guarantee, amounting to 150 million DMK, that was issued to LURGI. Viking Raps now had 550 million SEK, which was enough to start the two first phases of the project.

The aim was to start the procurement, the construction, and the operations in Lipetsk, and after a while, when Viking Raps could show a track record, to apply for more credit. The decision about purchasing for the first phase was to be made before the beginning of July 1991, and for the second and third phases in August–September 1991. The purchasing for the three phases amounted to 2 billion SEK. In May 1991, it was intended to put the plant into operation on January 1 1994, but the Swedes involved feared that the project would take much longer than the time that had been outlined. In March 1991, Karlshamns had sent a letter to some ten potential suppliers of equipment to Viking Raps. Karlshamns expected to receive definite offers by June 1991. Viking Raps selected one German and one American-British machine supplier. During the summer of 1991 Hermes and Kreditanstalt began to intimate that there would be problems with the credit guarantees. Finally, Hermes decided not to issue credit guarantees to Viking Raps. At the end of the summer of 1991 all the German money that was reserved for the Soviet Union was frozen. In

September 1992 no one knew with certainty from where the rape would come and how it would be paid for. In 1990, 90 000 tons of rape was harvested in the Lipetsk region, compared with the 1991 harvest, which was only 37 000 tons. There was neither sufficient will nor knowledge to deal with the harvest, although Karlshamns had tested the rape and it was of sufficient quality. After the German banks had frozen all funds, Viking Raps scaled down the project by two thirds. The end-products were no longer to be special fats and margarine but raw oil for export, and just one third of the production machinery would be erected. The need for credit thus diminished from 1 billion SEK to 250 million SEK. That would be enough to construct silos and to take the first steps in the production process. At the beginning of 1992 a new German fund was founded, with five billion DMK at its disposal, which was intended for the reconstruction of the food industry in the CIS. However, President Yeltsin did not accept the conditions set by the German government and the money was frozen once more until May 1992. Viking Raps maintained the contact with the German banks and suppliers, but the chances of getting credit from Germany were small. Instead, Viking Raps now mainly concentrated its efforts on convincing EBRD.

Period 3: 1991–1993

Market exit: During 1991 and the second half of 1992, the financial situation in Viking Raps became more critical. A letter of credit given by Karlshamns to Viking Raps expired on 30th September and now more money was urgently needed for the everyday running of the project. Karlshamns was not prepared to contribute with more credit. In September 1992, Karlshamns expected that a Swedish bank, a German bank, and EBRD, would give credit jointly. EKN had said that if another institution were prepared to participate, EKN would treat the application positively. In March–April 1992, the biggest shareholder, Lipetskpicheprom, ceased to exist. It had been broken down and was being privatized. Rospicheprom bought Lipetskpicheprom's shares. In the summer of 1992, Viking Raps' site was "occupied" by hundreds of people who were trying to stop the construction of the plant. Viking Raps had to leave the site, where the construction of the foundations for silos and installation of temporary electricity cables and gas pipes had begun, and restore it to its original condition. In the autumn of 1992, Karlshamns decided to investigate the consequences of the liquidation of Viking Raps and to work out a plan for liquidating Karlshamns' involvement in Viking Raps. Karlshamns did not have any hopes of finding the financial resources to realize the project. After September 1992, Karlshamns did not run any projected work. It was the Russian shareholders who financed the operations in Viking Raps. Mr Ivanov did not permit any use of economic reports or information, which meant that even Karlshamns and the board of Viking Raps had little insight into the operation of Viking Raps. In 1992, Karlshamns managed to convince Viking Raps to repay its debt to Karlshamns. During 1992 and 1993 Viking Raps met with various representatives from different authorities in Lipetsk and almost 20 sites were inspected, but that did not change Karlshamns' position and in October it liquidated its involvement in Viking Raps.

The case of Volvo in Iran

Period 1: 1973–1978

Positive expectation is increasing: Volvo started exporting to Iran in the 1950s. In 1964, after the Iranian government had imposed restrictions on the import of fabricated products, a joint venture contract for assembly production of trucks, tractors, and other agricultural machines was signed between Volvo and their former agent, Nasir, a wealthy Iranian businessman with political contacts and influence. The Volvo subsidiary, Bolinder Munktell (BM), had a leading position in the joint venture firm Zaymad. Nasir became general manager, and other managerial functions were undertaken by BM. In 1973, when oil prices rose, Volvo and Nasir became more optimistic about the future of the market. Nasir knew the managers of a firm called Dorman Diesel and, in 1974, Volvo started a licensing arrangement with Dorman Diesel for production of diesel engines. At the time, Nasir, together with two other rich and influential Iranians, established a new company named Rena Industrial Investment, which owned 75% of the shares in Zaymad. Volvo owned the rest. Zaymad itself owned shares in Dorman Diesel. Since the market was expected to become important to Volvo, their vice president was on the board of directors. The competitors were Mack, Mercedes, and British Leyland. Mercedes and Leyland had already established an assembly line similar to that of Volvo. Following the increasing sales, Volvo's expectation increased and therefore in 1976, there was a discussion between Nasir and Volvo's manager about a big new project with a production capacity of 20,000 trucks. The project was large when compared with the total production capacity of Volvo in 1976, which was about 28,000. Volvo could increase the export of components and spare parts to about 1 billion SEK. At the end of 1977, negotiations had progressed positively and the contract was to be signed in early 1978.

Period 2: 1977–1983

Negative expectation development: In early 1977, when the political demonstrations started, Volvo and local managers expected that the turbulence would soon stop. Nasir tried to provide Volvo with concrete information. In any case, Volvo's managers, both in Iran and in Sweden, were still positive about the future and had no other plans, as the market was considered very attractive and stable. When the political and economic situation became unstable in 1978, there was still pressure on Volvo from Nasir to continue negotiations. But the Swedish manager in Iran realized the increasing problems in production and state bureaucracy. He discussed the issue with headquarters and decided to slow down negotiations. In fact, Volvo was not affected by these political and economic problems because it was Swedish. After a short time, however, unexpected conditions arose. The strikes in Teheran were extended and started to disturb production. By the end of 1978, the government froze all foreign exchange transactions, and Volvo could not send export payments from Iran. The increasing turbulence stopped production and forced the manager to send Swedish personnel back to Sweden. The manager himself stayed on to watch the political development.

Volvo's managers were still positive about the future and believed that their production would not be affected, even if a new political group took over. However, they did not know how long the disturbances would continue. The production lines of competing firms were closed down and their managers had already left Iran after the Shah's departure and the proclamation of the nationalization of foreign MNCs. Volvo's major problem was having Nasir – with his relationship with the Shah – as a partner. The new leadership showed that they did not have confidence in Volvo's operation because of the partners. In 1979, Volvo suddenly received a message that the truck factory in Iran would be nationalized. Volvo's 19% share was to be taken over by the government. After several meetings, Volvo headquarters decided to keep the manager in Iran, but it was more costly to stay than to leave. Although production was closed down, the company still had more than 1,400 employees on its books.

The revision group from the government studied all the details in the contracts, and the future of Volvo became uncertain. After several months of investigations, the revision group gave a positive report on Volvo and determined that Volvo could keep its ownership. The Ministry of Commerce took over the responsibility for the Rena, Zaymad, and Dorman companies, and replaced the boards of directors of the three firms. After a short period, the responsibility was transferred to the state organization IDRO, which selected new members for the boards of directors. The new members were chosen because of their religious backgrounds and lacked technical knowledge. But the problem was not the background of the directors: it was rather the continued political turbulence, suspicions among local managers, the bureaucratic system, the war, and Volvo's being left alone without knowing what to do.

During the investigation by the revision group, Volvo stopped all deliveries to Iran. IDRO made several efforts to persuade Volvo to be included on the board of directors in the joint venture firms. The reason was to show Volvo's confidence, as the Volvo manager, unlike those of its competitors, had stayed in Iran. Besides, IDRO had realized its technological dependence on Volvo. In the meeting arranged in 1982 by the Swedish Export Advisory Board, MHI and IDRO asked Volvo to discuss existing co-operation problems. They asked Volvo to take a position on the board of directors, but Volvo refused. Despite all the problems, the Volvo manager had a positive future view of the Iranian market; without this, Volvo would already have left the country. The former market commitment had at least brought Volvo closer to the buyers. By the end of 1982, a new manager in Volvo became responsible for the Iranian market.

Period 3: 1983–1993

Escalation of market commitment: In 1983, Volvo had an office involving two specialists and a few technicians to control local production. IDRO protested against this arrangement, which Volvo considered necessary because its name was on the products. The real reason was to watch the market. At that time, foreign firms were not allowed to have representatives in Iran. The strategy was no longer related to the earlier investment, because Volvo had already given up its ownership to the local authorities. IDRO again requested Volvo to include a manager on the board of directors, but Volvo refused. This waiting strategy seemed to have positive consequences and the manager succeeded in signing contracts for three minor projects: (a) truck motors; (b) components for the trucks; and (c) motors for generator stations.

After the consolidation in 1984, the level of personnel turnover in the business and political systems – the two IDRO managers kept their positions for seven to eight years – gradually decreased, which provided the stability needed for co-operation between IDRO and Volvo. Increasing stability affected Volvo's real views and managers started to believe in market progress in the future. In 1984, there was a discussion about another importing and assembly co-operative venture with Volvo. Increasing stability increased Volvo's belief in the future. The negotiation led to co-operation with the firm Kaveh, which had earlier belonged to a US company, Mack. The contract concerned selling parts and did not require any financial investment. The two firms Zaymad and Kaveh, which dominated the market with a production capacity of 2,000 trucks each, received technical assistance, including spare parts and service, from Volvo. Until 1992, Volvo's manager did not accept any official positions in Rena, Zaymad, or Dorman, but received assistance from them when tendering offers for new projects. This led to the selection of Volvo for some projects. From 1986 to 1988, the last years of the war, the sales level was very low, but in 1990, Volvo was the dominant firm in the truck business. After 1990, besides old competitors, some new competitors (e.g., Iveco) had become active. The new competitors penetrated the market by offering low prices. Despite increasing competition, however, Volvo maintained a strong position.

Discussion

Three kinds of expectation

The two cases manifest particular circumstances, which yield a clear picture of the expectation, knowledge, commitment decisions and current activities. The traditional use of the IP-model emphasizes an incremental process and does not anticipate or explicitly integrate the factors that can explain the behaviour in conditions where the cumulative process falls short. The case of Volvo in the first period shows that as far as the development of knowledge and experience are incremental and changes are smooth, the construct of experiential learning is a sufficient base to explain commitment decisions. In this case, the future dimension of a commitment decision was woven in and built on past learning. The factor of expectation was thus concealed behind experiential knowledge. But, as shown in the last phases of these two cases, when changes are no longer smooth and predictable the traditional exploration of the concepts is inadequate to explain the commitment decisions. Drastic changes interrupt the balanced cumulative progress in commitment and knowledge. Consequently, past learning is unable to ground the commitment decisions and so the factor of expectation clearly appears as an explanatory variable.

In this study we follow the definition of Simon (1976) and define it as values given to the future constructed by hopes and knowledge. More explicitly, expectation is the firm's defined probabilities of the occurrence of positive and negative events if the firm should engage, or aims to engage, in some interaction with others in a market (Hunt 1991; Oliver 1980). For the cases, revealed expectation changed in both directions. The cases show both the incremental progress in the firms' behavior when they escalate their commitments in the first phases and also the negative development in the expectation. The purpose of the process analysis of the cases is to explore how the two firms' commitment decisions changed in one direction or the other because of changes in the expectation. In the first phase of the case of Volvo, there is successive incremental development. In the same phase, the case of Karlshamns describes a rapid internationalization by joint venture. The case of Volvo in the last phases illustrates the third type of development, which is reinternationalization. The fourth and final type is retrenchment and exit, as is seen in the case of

Karlshamns. In sum, expectation first developed positively in both cases but took a negative direction in the second period. The negative expectation in period three led to Karlshamns deciding to abandon its commitments and exit, but Volvo, which still had a positive expectation, stayed and decided to commit new resources. The two cases clearly manifest different types of expectation developments, which result in different directions in the internationalization processes. A way to describe developments in these cases is to make a distinction between different types of expectation, which describes both cumulative and digressive commitment decisions. The cases show three distinct types of expectation: general, relationship, and network. It is vital to underline that these three types of expectation often exist simultaneously, but they are weaker or stronger, depending on the situation in the market. In a sense, general expectation gives rise to relationship expectation, which leads to network expectation. However, the cases indicate two crucial aspects of the relation between these expectation types and commitment decisions and experiential knowledge. One aspect concerns the balance between the development of the expectation and the variables of commitment and experiential knowledge (i.e., an incremental process in expectation). The second aspect is the imbalance between the expectation and experiential knowledge.

Balance between the expectation, commitment, and experiential knowledge

For the first aspect, the discussions in the cases, specifically in the first phase of the Volvo case, illustrate the fact that expectation, similar to other factors in the IP-model, is progressive. That is to say, that the firms decide to make a commitment as their general expectation is positive. General expectation (Gabarro, 1978; Lindskold, 1978; Sitkin and Roth, 1993) is engendered when an actor gives common and homogeneous values to the groups of market and institutional actors or factors and their future development. General expectation, which manifested in the case of Karlshamns in the first phase, and Volvo in the second phase (when Volvo had lost all its commitment and links to market actors), initiated the process of building relationships. The more specific knowledge gained constructed the firms' expectation towards particular market actors. In the first period, Karlshamns approached into the market despite the fact that it had no experiential knowledge of that market. General information and facts interwoven with wishes indicated positive progress in the Russian economy. Karlshamns' commitment decision for penetration was driven by the managers' positive general expectation.

As the firm interacts with some actors in the market, its knowledge becomes richer and contains more details. The firm's values become more context specific. The expectation values given to actors become heterogeneous which give raise to the relationship expectation. This means that the firm evaluates what it can expect from each actor if it becomes committed. By increasing knowledge about business firms in Russia, Karlshamns' managers believed that one of the local actors would be appropriate as a licensing partner. However, despite the fact that Karlshamns' managers knew about the risk and had plans not to become highly committed in the market, they signed a large joint venture contract. As the two partners did not have a long history of co-operation, the expectation of the firm in this relationship was not driven by experiential knowledge. Furthermore, it was interwoven with the wish that the local partner would have the ability to solve local problems. Karlshamns also established relationships with foreign actors to gain approval for the needed credit or other business activities. Network expectation goes beyond the direct dyadic relationships; however, it still concerns specific actors, but they are part of the firm's network via connected relationships. As far as the knowledge of a firm beyond the dyadic relationship is general, the expectation contains the two components of relationship and general. Increased knowledge about connected or potential actors gives network expectation. It is specific values given by the firm to both the direct and connected actors. The network expectation contains heterogeneity and is first of all, actor-specific.

Volvo's increasing commitment decision in the first phase did not rely only on general or relationship expectations. The firm had already been in the market for a long time and had well-established relationships with local actors when facts about the increased economic prosperity in the market were released. Since Volvo had a positive expectation of its former partner, it increased its commitment through a joint venture. In the first phase of Volvo's activities in Iran, the commitment decisions were developed on the base of market learning, which included facts about what the firm expected to gain from the increasing economic prosperity in Iran. This development increased the incremental growth in expectation, which concerns the balance between the development of the expectation and the variables of commitment decisions and experiential knowledge. In this phase, Volvo's commitment decisions were developed together with experiential knowledge. But, in the Karlshamns case, in the first phase, it was the positive general expectation of the market and political performance that made the firm enter that market. The firm had no specific knowledge about specific actors in Russia. After penetration, the expectation that the partner in Russia would fulfil its obligation in the relationship pushed forward a joint venture contract. Karlshamns' expectation relied on the fact that the local partner would handle his task properly. Additional knowledge about Russian market actors extended the expectation from a relationship expectation towards a network expectation. Network expectation is predicated on the behavior of actors in the network. As much as the market commitment and knowledge of the focal actor toward each actor is dissimilar, the expectation towards each actor becomes specific. The evolution of expectation from relationship to network, as illustrated in the case of Karlshamns in the first phase and Volvo in the second and third phases, begins with general expectation and develops into relationship and network expectation. The cases also expose another fact about expectation types. One major reason for the joint venture in Iran was the Volvo manager's general expectation (i.e., a belief in general economic prosperity in the future). Thus, a highly committed firm has all three types of expectations. Firms, when making decisions about new commitments, will assess the socio-economic development, their close partners, as well as all those connected with the project. This further elucidates the fact that relationship expectation contains general expectation. Karlshamns, when deciding on co-operation with its partner in Russia, also had a positive expectation about the behavior of the institutional actors.

Imbalance between expectation and experiential knowledge

The second aspect, the imbalance between the expectation and commitment decisions and learning, was exposed for Volvo in the second phase and for Karlshamns in the second and third phases. Such a development can occur with the instability and unpredictability of the situation, and can, for example, interrupt the smooth growth in commitment and make the firms decide whether to stop or to escalate market commitment (Staw, 1981). The imbalance can also arise in conditions when a decision regarding for example, market penetration, is not grounded in market learning, but instead, is based on a positive general market expectation. The expectation and commitment decisions of these two firms in different time periods is summarized in Table 9.1. As shown in that table, the driving force for Karlshamns to escalate commitments in the Russian market was the positive market expectation and not the stepwise learning. In the second case, Volvo had a long period of experience in the market but did not increase its commitments until 1974. As mentioned above, it was the increasing oil income and the expected prosperity in the country that made the firm and its partner develop a positive belief in the future. At the end of this period, Volvo had established relationships with several local actors and knew that its local partner would undertake the main local tasks. All three types of expectation were cumulative and governed the commitment decisions in this network. Contrary to the first case, the joint

TABLE 9.1

A summary of the expectation and the commitment decisions of the two firms

Firm	Phase	Expectations	Commitment decisions
Volvo	1	A high expectation for all types.	Incremental learning process. Commitment decisions for several joint ventures.
	2	High general expectation, no relationship, or network expectation.	Loss of knowledge and commitment. Decision to stay and escalate commitment.
	3	High general expectation, relationship expectation is conceived, building network expectation.	Increasing knowledge, and commitment. Decision to escalate commitment with joint venture.
Karlshamns	1	High general expectation and high relationship expectation.	Decision for joint venture commitment. Low general, relationship, or network knowledge.
	2	At the beginning high general, relationship and network expectations, followed by a reduction in the general expectation.	Decision to increase the commitment in the beginning, but later, at the end of this phase, managers decided to reduce financial inputs.
	3	The process of increasing negative expectation at all levels continued and finally the firm had no positive expectation left about the Russian market.	As their learning about the market increased, the firm eventually realized that it did not know much about this market and the future possible changes. The knowledge they obtained gave a negative perception about the future and the only decision alternative was to abandon all commitment and to exit from the market.

venture in the Volvo case followed the cumulative process in the IP-model. In the Karlshamns case, the managers' decision concerning the firm's commitment was based, firstly, on their general expectation. Their joint venture commitment decision, made shortly afterwards, was driven by relationship expectations and not relationship knowledge. Commitment decisions and expectations developed in parallel, but remained in imbalance with the experiential knowledge. Managers' expectations that their Russian partner would handle the local issues and that the general knowledge was sufficient for a large joint venture commitment, after a short period of relationship development, indicates that wishes were blended in with the knowledge on which the expectation was based (*cf.*, Cyert and March, 1963).

At the beginning of the second period, Karlshamns had high general, relationship, and network expectations. Karlshamns expected that its partner would solve the critical business issues and that the political actors in Russia would act for business prosperity. These beliefs gave the signal for further commitment. Inability of the local partner, together with improper political decisions, led to critical problems for Karlshamns. The connection between the political actors in Germany and Russia made the banks withdraw credit from the joint venture and Karlshamns was left with commitments and no knowledge about the future. The only driving factor was their diffuse expectation that the market would change. A similar, although more dramatic, condition was faced by Volvo. At the beginning of the crisis, all market actors expected a short period of drama. No one knew what would happen. Despite this fact, all had a vague but positive picture about the future. That kept Volvo in the market. When new market actors took over the business and political positions, Volvo's high commitment, based on earlier learning and relationships, became useless as a basis for decisions. The network dissolved and Volvo was left alone in a hostile environment with neither market nor general knowledge. The loss of commitment and the change in the local network had worsened the situation. Information about local conditions was contradictory and unreliable, but the critical decision of whether to exit from the market or to stay had to be made. All signs indicated that exit was the rational strategy. Through the transfer of general knowledge and learning from other markets – that the situation would improve – Volvo built a positive general picture for the future of the market in Iran. What this means is that Volvo lacked relationship and network expectations but decided to stay and commit itself to the market because their general expectation was positive. However, Volvo and Karlshamns, at the end of phase two, had an imbalance in the three variables of commitment, knowledge, and expectation.

For Karlshamns the development of negative expectations continued in period three. In fact, one of the critical problems for Karlshamns was that of financial credit. Being connected to the Russian political actors via credit providers made the joint venture firm vulnerable to political decisions. With the development of negative expectations, Karlshamns decided to reduce its commitment and to cut down one third of the production capacity. The problems accelerated with the unexpected occupation of the facilities and privatization of the local partner. Karlshamns' managers realized that the local firms, local partner, and institutions in Russia were trapped in a negative development direction. Hence, all three types of expectations developed negatively, together with their market learning. Accordingly, even when the Russian shareholder started to finance the project, and despite the fact that they were highly committed in the market, Karlshamns decided to exit from it. For Volvo, on the contrary, the market had almost returned to its original state only because of the positive general expectation. The firm had no relevant market knowledge, but the local political and market actors had a general understanding that the firm, despite all risks and costs, had been honest and stayed in the market. Subsequently, the increasing general expectation led Volvo to increase its market commitment cautiously. At the end of the 1980s, when the position of the local actors became more stable and the relationships between local actors and Volvo were extended, Volvo decided to engage in bigger projects and to discuss joint venture contracts in the 1990s. Where Karlshamns managed their reduced expectation with retrenchment followed by exit from the market, Volvo selected re-internationalization and began to increase its commitment through a process of gradual learning and expectation. Absence of knowledge will naturally make the expectation move from one of relationship or network to what is called here a dispersing phase. This is reverse dynamism in the process of expectation change, from high expectation to despair. As far as the evolution of expectation development is progressive, drastic change can produce sudden change in the expectation. Critical conditions in the activities of connected actors or in a dyadic relationship affect the probabilities of future outcomes and can invert and disperse expectations, which can lead to the withdrawal from the market.

Expectation – the missing link in the IP-model

In both conditions of penetration leading to expansion, or a decrease in commitment leading to exit, commitment decisions are grounded in the expectation of the firm. As discussed in the case analysis and shown in Figure 9.1 below, the nature of commitment decisions depends on what a firm expects from its general environment (i.e., its general expectation); its relationship partner's behavior (i.e., its relationship expectation); and the attitude of the actors in its network (i.e., its network expectation). Thus decision commitments rely on the cumulative content (positive or negative) of all these expectations together.

In the following, the results are summarized to provide a precise clarification of the concepts discussed in the earlier sections. The facts have led us to construct a conceptual frame that incorporates and connects the different concepts interwoven in the case analysis. As depicted in Figure 9.1 below, the frame elevates the concept of expectation and locates it in between the concepts of commitment decisions and experiential knowledge and exogenous factors.

Exogenous factors and experiential knowledge refer to political and market knowledge, which may be external or internal to the firms. As elucidated in the cases, they affect the types of expectations and

FIGURE 9.1
Expectation in the IP-Model

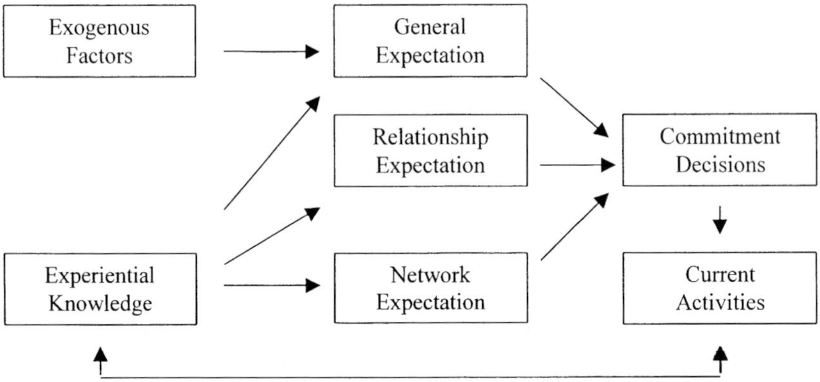

subsequently the commitment decision which affects the current activities and experiential knowledge. Exogenous factors refer to information on facts, speculations, or reputations that a firm collects by either observation or pure market studies. It can be seen as analogous to the firm's wishes, which Cyert and March (1963) present, or to microeconomic or marketing studies on expectation and market analysis. Experiential knowledge is obtained through interacting and mutual exchange relationships, as demonstrated in the case of Volvo in the first and third periods. The exogenous factor denotes information collected from external sources to build an expectation and possible commitment decision. Karlshamns' market entry and even their joint venture decision was completely based on exogenous factors. As far as the experiential knowledge is firm-specific and based in the exchange relationships, exogenous factors are expostulations that rely on facts or speculations that, with simple market efforts, can be available for all firms. Experiential knowledge represents accumulated knowledge on relationship-partners and actors in the network.

Accordingly, incrementality in commitment decisions relies on a positive development of all three types of expectations set by positive market experiences. It relies on a coherence or balance in the development of experiential knowledge, expectation, commitment decision and current activity factors. This development can be incremental, when experiential knowledge, expectation, commitment decision and current activity factors advance positively. Alternatively, it can be developed in a contrary direction; in other words, incrementality for all three mentioned factors can embody a negative nature. The first phase in the case of Volvo provides evidence of a positive incremental development. Volvo's first low level of export commitment started with a positive general expectation and advanced through the development of expectations concerning relationships and network based on increasing experiential knowledge. Beside the above two conditions of positive (the first phase in the Volvo and Karlshamns cases) and negative development (the second and third phases in the Karlshamns and second phase in the Volvo) processes there are cases that indicate an incoherence in the development of these three factors. The case of Karlshamns can be used to illustrate this third condition. In this case, the dominant factor for penetration and a joint venture contract was based on an exogenous factor. The firm expected that conditions would improve since the general reports and newspapers provided such information. It also expected that its partner would act according to the contract and 'wished' it so although there had not been any earlier experiences of co-operation. The high expectation made the firm decide in favour of a high market commitment, despite the fact that experiential knowledge was low. Incoherence between expectation and experiential knowledge affects the properties of the commitment decisions.

Conclusion

The paper uses the facts to generate some ideas on expectations in the IP-model. Adding the concept of expectation may provide some response to those who criticize the IP-model for its inability to explain the processes of entry, retrenchment, or exit. The evidence of this paper not only demonstrates the positive cumulative process in internationalization, but also, and mainly, emphasizes the negative development and the reasons for that. One of the critical questions, for example, was how to explain behavior in conditions such as when firms, despite their low or obsolete market knowledge, continue to exist in the market or even to escalate their commitment. The variable of expectation gives the clue to understanding commitment decisions, not only for cases in which foreign MNCs are operating in unpleasant environments, but also in cases with stable environmental conditions. For analysis of those cases that follow an evolutionary process, incorporation of the expectation concept into the IP-model may seem excessive. It may be that knowledge and commitment follow a balanced progression and the concepts are sufficient to explain the behavior. However, the concept of expectation can provide more concrete aids to improve our understanding of behavior outside the evolutionary process, such as rapid penetration with large investment in a foreign market.

Defining expectations with the IP-model, the evolution of expectation from relationship to network relies not only on the development of knowledge by one specific partner but also on expectations on the other related actors. Based on experience and commitment, firms can predict the reactions of the related actors (Barber, 1983; Rotter, 1980; Shapiro, 1987). The three types of expectations indicate this incremental process in this study. They are general expectation, which gives rise to relationship expectation, which, in turn, leads to network expectation. The idea of an incremental process also promotes another particular aspect: that the network expectation could be expected to contain the other two types, and that the relationship expectation needs to contain the general type of expectation. Foreign firms holding all types of positive expectation are highly committed firms.

Indeed, the cases show that a cumulative process of internationalization is built on a progression in all three variables: (a) experiential knowledge and exogenous factors; (b) expectation; and (c) commitment decisions; and (d) current activities. Market knowledge is a subjective phenomenon and changes need to be large so that the new experiences and information can be accommodated within the cognitive framework. A change in one actor's behavior may result in changes in the experiential knowledge, expectation, and consequent commitment decision of another actor. The evolution of expectation, commitment, and experiential knowledge factors is accommodated with smooth changes, and progress in all three requires a balanced structure. Sometimes, new information and experience cannot be integrated into the existing framework, which leads to the collapse of the framework. Firms seek new interpretations, such as giving a higher value to expectations that are based on wishes, or transferring knowledge from other networks, when they make decisions to escalate their commitments.

Commitment decisions in the evolutionary process, which evolves from progress in the cognitive perceptions of what has happened (encompassing experiential knowledge) and what is supposed to happen (that is, expectation) can sometimes extend to problematic situations. One variable can be perceived to have a positive and another a negative development. This interpretation explains the behavior in conditions where knowledge is obsolete but commitments escalate. Such problematic conditions denote an imbalance between the experiential knowledge and exogenous factors and committed resources, and raise the significance of expectation. Negative development in all three variables (expectations, experiential knowledge and exogenous factors) leaves the option of exit. However, inquiry into such behavior reveals that all types of expectations can have a negative development. Under conditions where two types of expectations develop negatively, but where there is positive progress in one type, the decision involves a choice of escalation of commitment or reinternationalization. In conditions where expectations develop negatively for a long period, a decision to exit from the market becomes the feasible alternative. Complexity increases when these expectations do not develop along the same track, specifically, when expectation – knowledge – transfers from one market or relationship to another. General knowledge is the common type of knowledge that transfers from one market to another. Experience from one market that transfers to another affects the expectation of a specific market. In a market with negative current development, a firm can give positive value to its general expectation not because of a positive development of expectation in that specific country, but because of the transfer of knowledge from other markets.

References

AXELSSON, B., and JOHANSON, J., 1992, Foreign Market Entry – the Textbook vs. The Network View, in Axelsson, B., and Easton, G., (eds), *Industrial Networks – A New View of Reality*, pp. 218–234, London: Routledge.

BARBER, B., 1983, *The Logic and Limits of Trust*, New Brunswick: Rutgers University Press.

BLANKENBURG-HOLM, D., and JOHANSON, J., 1992, Managing Network Connections in International Business *Scandinavian International Business Review*, Vol. 1,1, pp. 5–19.

CYERT, R. M., and MARCH, J. G., 1963, *The Behavioral Theory of the Firm*. 2nd edition, Cambridge, Massachusetts: Blackwell Publishers.

FINK, S., 1986., *Crisis Management*, New York: American Association of Management.

GABARRO, J. J., 1978, The Development of Trust, Influence and Expectations, in Athos, A.G., and Gabarro, J.J. (eds), *Interpersonal Behavior: Communication and Understanding in Relationships*, pp. 290–303, Englewood Cliffs, NJ: Prentice Hall.

HADJIKHANI, A., 1996, *International Business and Political Crisis*, Uppsala: Almqvist and Wiksell.

HADJIKHANI, A., and JOHANSON, M., 1996, Facing Foreign Market Turbulence: Three Swedish Multinationals in Iran *Journal of International Marketing*, 4 (4), pp. 53–73.

HUNT, S.D., 1991, *Modern Marketing Theory: Critical Issues in the Philosophy of Marketing Science*, Cincinnati: South-Western.

JOHANSON, J., and MATTSSON, L-G., 1988, Internationalisation in Industrial Systems – A Network Approach, in Hood, N., and Vahlne, J-E. (eds), *Strategies in Global Competition*, pp. 287–314, London: Croom Helm.

JOHANSON, J., and VAHLNE, J-E., 1977, The Internationalization Process of Firm – A Model of Knowledge Development and Increasing Foreign Market Commitments *Journal of International Business Studies*, 8, Spring/Summer, pp. 23–32.

JOHANSON, J., and VAHLNE, J-E., 1990, The Mechanism of Internationalization *International Marketing Review*, 7 (4), pp. 11–24.

JOHANSON, J., and WIEDERSHEIM-PAUL, F., 1975, The Internationalization of the Firm – Four Swedish Cases *Journal of Management Studies*, 3, pp. 305–322.

KAUZMAN, A., and JARMAN, A., 1992, Creeping Crisis, Environmental Agendas and Export Systems. A research note paper delivered at the 22nd International Congress of the Institute of Administrative Sciences, 13–17 July, Vienna.

LINDSKOLD, S., 1978, Trust Development, the GRIT Proposal, and the Effects of Conciliatory Acts on Conflict and Cooperation *Psychological Bulletin* 85 (4), pp. 772–793.

MAKHIJA, M. V., 1993, Government Intervention in the Venezuelan Petroleum Industry: An Empirical Investigation of Political Risk *Journal of International Business Studies*, 24, No. 3: pp. 531–55.

NILSON, T., 1995, *Chaos Marketing, How to Win in a Turbulent World*, London: McGraw-Hill.

Amjad Hadjikhani and Martin Johanson

OLIVER, R. L., 1980, A Cognitive Model of the Antecedents and Consequences of Satisfaction Decisions *Journal of Marketing Research* 17 (November), pp. 460–469.

ROTTER, J. B., 1980, Interpersonal Trust, Trustworthiness, and Gullibility *American Psychologist*, 35, 1, pp. 1–7.

SHAPIRO, S. P., 1987, The Social Control of Interpersonal Trust *American Journal of Sociology*, 93, 3, pp. 623–658.

SIMON, H., 1976, *Administrative Behaviour, A Study of Decision Making Processes in Administrative Organization*, New York: Macmillan.

SITKIN, S., and ROTH, N., 1993, Explaining the Limited Effectiveness of Learning Remedies for Trust/Mistrust, *Organization Science* 4 (3), August, pp. 367–392.

STAW, B., 1981, The Escalation of Commitment to a Course of Action *Academy of Management Review*, 6 (4), pp. 577–87.

CHAPTER 10

Creation and Diffusion of Knowledge in Subsidiary Business Networks

CECILIA PAHLBERG

Introduction

As stressed by a number of researchers, the ability to innovate is an important source of competitive success (see e.g. Bartlett and Ghoshal, 1989, p. 115). Hence, it is essential to be ahead of the competitors in the process of creating knowledge. This is of considerable interest in large and complex organizations such as multinational companies (MNCs). A central theme in the modern literature on MNCs is the discussion about worldwide learning, i.e., how knowledge should be diffused among units/subsidiaries. However, there is less discussion about how knowledge is developed. Birkinshaw (1997) states that many MNCs seem to neglect the creative potential of their subsidiaries and that initiative[1] at the subsidiary level is an under-researched phenomenon. According to him, "subsidiary initiative has the potential to enhance local responsiveness, worldwide learning and global integration" (p. 208).

In this chapter the focus will be on the role of the subsidiary in the creation of knowledge. Today, the view that MNCs are heterogeneous entities consisting of subsidiaries with different capabilities and competencies seems to be widely accepted. As stressed by, for instance, Gupta and Govindarajan (1991, 1994) and Bartlett and Ghoshal (1989), subsidiaries

---

[1]His definition of initiative (p. 207), "a discrete, proactive undertaking that advances a new way for the corporation to use or expand its resources", goes back to Kanter, 1982 and Miller, 1983.

Cecilia Pahlberg

have different roles, such as being innovators or implementors. But how do they get these roles? The most common perspective is that the subsidiary's role is determined by headquarters and assigned to the subsidiary, i.e., subsidiaries are *given* their role as innovators, implementors, etc. Another perspective is that it is the importance of activities at subsidiary level that determines the role. Relationships with specific customers and suppliers and other important counterparts in the subsidiaries' networks are of interest if one is to understand why some subsidiaries become more important than others. But it is also essential to take relationships within the MNC into consideration, such as those the subsidiaries have with sister-units and other units in the corporate system. This view of the MNC as an interorganisational network with relationships to actors both inside and outside the formal boundaries of the MNC (Ghoshal and Bartlett, 1990, Forsgren and Johanson, 1992) is gaining popularity amongst researchers.

Learning occurs in many different functions within a firm, but the focus is often on R&D. As, for example, von Hippel (1988) and Håkansson (1987) have shown, learning from external counterparts, such as customers, is essential, particularly where innovation is concerned. However, as Dodgson (p. 389) stressed: "given the *centrality* of R&D as an organizational learning mechanism, there is a surprising paucity of research across all traditions on the broad question of learning in R&D, its promotion and funding, and the subsequent *diffusion* of learning throughout the organization." While R&D is an important source of learning, learning also takes place within other functions such as manufacturing and marketing. However, wherever learning takes place, much of the existing analysis is limited to its outcomes, and ignores or underestimates the problems and complexities involved in the process of learning. In the following description, the aim is to illustrate both the creation and the diffusion of knowledge within an MNC.

Organizational learning is nowadays the focus of considerable attention. Since 1991 when *Organization Science* had a special edition on "Organizational Learning", interest in this subject has been pronounced, but the field can be traced as far back as to Weber. Research on learning has a long tradition, especially within the field of psychology where the focus is on the individual. Research on organizational learning is a much more fragmented and multi-disciplinary phenomenon. It involves a number of academic disciplines, such as economic history, industrial economy and management studies. Amongst the reasons why this theme is so fashionable right now, the following can be mentioned (see Dodgson, 1993):

– It is increasingly stressed that learning is a key to competitiveness;
– The rapid pace of technological change forces firms to do things in new, and often radically different, ways;

170

- The learning concept is dynamic and emphasises the continually changing nature of organizations. It is also integrative as it can unify different levels of analysis (individual, group and corporate level).

A concept closely related to learning is knowledge. In this paper knowledge is regarded as the outcome of learning, and following the arguments by Weick (1991) made with reference to Duncan and Weiss (1979), learning is seen as "the process within the organization by which knowledge about action-outcome relationships and the effect of the environment on these relationships is developed." Hence, learning involves the development of knowledge about relationships.

The relationship between learning by individuals and collective learning has also received much attention. The starting-point here is that individuals learn in performing their daily activities and it is their knowledge that shapes the organization's collective memory in the form of the routines and procedures instigated. The collective memory is a result of present and preceding activities, and whilst it is individuals that form it, it also shapes their learning, i.e., it may both enable and restrict the actions of individuals. Above it has been stressed that individuals learn from their daily business activities. The exchange with a counterpart, with a customer, for instance, will often lead to increasing commitment, adaptation and interdependence – the business relationship develops. As knowledge is often tacit, learning-by-doing is essential, i.e., the individuals involved in the relationship gain experience and knowledge through working together. The individuals modify their activities incrementally; in turn this modifies the collective memories of the two relationship partners in such a way that the joint productivity of the firms is raised (Eriksson *et al.* 1998).

Knowledge has emerged as "the most strategically-significant resource of the firm" and integration of knowledge is a firm's most important task (Grant, 1996, p. 375). Although much has been written about knowledge, there is still a need for qualitative and quantitative research in this area (Miner and Mezias, 1996). The description below is a qualitative contribution illustrating how knowledge is developed in a foreign subsidiary within a Swedish MNC as a result of interaction with specific actors in the local network, and how this knowledge is diffused within the subsidiary and within the MNC as a whole. It is based on several interviews with the managing director and some of his colleagues in the subsidiary being examined, with other subsidiary managers within the MNC and with headquarters.[2]

---

[2]The company has been visited several times during the period 1991–1998. The interviews have been both structured with extensive questionnaires and tape-recorded open discussions.

It will be illustrated how a subsidiary's business relationships are a crucial source of learning and of capability development. Through its business relationships with specific counterparts in its business network, a firm creates unique knowledge. But it is not only the activities performed within a relationship that are important, firms also learn through co-ordinating their activities with those performed in connected relationships and by generalising their activities to other relationships (see Eriksson *et al.*, 1998). Hence, the description of the creation of knowledge will be divided into three parts: learning in specific business relationships, co-ordinating activities and generalisation. In a subsequent part, the diffusion of knowledge within the MNC will be elaborated upon.

Knowledge creation in a business network – the case of Powerpac

The company

The company, which will be referred to as Powerpac, was established at the beginning of the 1950s for the purpose of producing solid board. Some years later it started producing corrugated packaging, and today that is the main product of the company. Powerpac expanded rapidly in the 1960s and 1970s and was acquired by a Swedish MNC in 1976. The policy of the new owner was not to intrude upon the subsidiary's business, and the firm was managed with great autonomy until the MNC merged with another Swedish company in the mid 1990s. Today, Powerpac consists of nine domestic companies/production units and three foreign-based sales and production units. There are 1 100 employees, of whom 10 per cent are abroad, and the turnover is above 1 billion SEK. The market share in the country is just above 30 per cent (50 per cent attributable to simple, brown boxes, and 50 per cent to refined products), which makes Powerpac the country's leading producer of corrugated boxes. Its three main competitors have a market share of between 20 and 30 per cent apiece. Hence, the market can be characterised as being oligopolistic and the main possibility for a company to increase its turnover is through the development of new products.

From the outset, Powerpac has been eager to seek new methods and techniques and, according to the managing director, "curiosity" and "education" have always been key words in the company. To increase knowledge is seen as a key challenge and it is also stressed that the company wants "to lead the way, to be ahead of development". As they say, "With our unique ability to combine creative thought with fast action, we are certain to remain one of the leading packaging companies in Northern Europe." But how do they achieve this? Where does the company find its input for development?

The following section elaborates upon how product and production processes are developed within the subsidiary. According to the managing director, there is a very good supply of competent people in the environment. There is also easy access to raw material, although the paper mostly comes from foreign paper suppliers. But, according to the managing director, the two aspects that are of particular importance for the development capacity are the situation within Powerpac and its relationship with specific customers.

Learning in specific business relationships

In the introduction it was indicated that learning mainly takes place when individuals perform activities in business relationships. In this case, the relationship with two customers, Playo and Lux, seems to be most important. These customers have different demands: while the printing quality and the design of the boxes are most important for Playo, Lux is mainly interested in the strength of the packaging.

In the packaging industry, it is typical that the customers are located within a radius of 200 km from the plants as transport costs would be too high otherwise. The number of customers is generally also considerable – for Powerpac it exceeds 4000 – and usually no single customer accounts for more than a few per cent of the turnover. In this particular instance Lux is accountable for an unusually high percentage – almost ten per cent – while Playo is considerably smaller.

Although Playo is not the largest customer, it is considered to be the most important one. It is a world-famous toy-producer and its relationship with Powerpac is about 30 years old. According to the people at Powerpac it is considered to be "impossible to replace". At the start, Playo bought ordinary brown boxes for storing and transporting toys. For Playo, receiving regular deliveries from Powerpac, once a week or even more often, was the most important aspect. When these deliveries failed at the beginning of the 1980s, the relationship was broken off completely, and it took more than five years for Powerpac to regain Playo's confidence. They managed to do this by approaching Playo's R&D department, and, together with their engineers, developing a new box of vital importance for the marketing of Playo's products. For their product, the design of the box as well as the print on it is essential as such features attract the interest of the customers. Hence, Playo wanted boxes with excellent print, as the boxes themselves were important for the marketing of the product. So when the people from Playo and Powerpac met, they concentrated on the colours on the packaging and the co-operation furthered the development of pre-print liner, which gives a quality close to offset and provides print in six colours. This enabled Playo to replace their formerly used plastic boxes with boxes made of paper. The managing director at Powerpac argued for investment in a plant for production of pre-print liner and the management in Stockholm supported him as such an investment could be of help for other units within the Group. The investment was approved and a new plant, the first in this part of Europe to produce pre-print liner, was established.

In the mid 1980s, the people in Playo's R&D department were able to convince their purchasing department that Powerpac could be regarded as a trustworthy supplier again. Since then the relationship has been strengthened further and it is now considered to be too strong to be broken. A number of people from the two companies visit one another regularly and 15–20 people from the sales/purchasing and R&D departments are involved in these direct contacts. Today, the most important project with Playo is to develop smaller, more compact packaging. Playo used to have large packaging in order to motivate the high price of their products, but the influence of the environmental movement on end customers has resulted in customers putting strong demands on Playo, expecting their products not to be injurious to the environment. So Playo now puts all its efforts into avoiding negative publicity, such as being the recipient of the "waste-award".

Thus, Playo's main importance is for Powerpac's technological development. As a very demanding customer, it has a considerable effect on the product and production development in Powerpac. But most important, it is through the co-operation that Powerpac has realised the importance of design and of quality print. As one of the respondents said: "The relationship with Playo has given Powerpac knowledge in graphic design, which is of use in the whole company." Not only is Playo important for the technological development in Powerpac, it is also a source of information about market activities and governmental issues. Furthermore, as a company with a very high reputation, it serves as a bridge to other organizations and new business contacts.

The importance of Playo for Powerpac can be seen by the fact that Powerpac has not only adapted its product and production technology to meet Playo's demands, but it has also changed how it conducts its business and, to some extent, even its organizational structure. Through these adaptations, the activities of the two firms are co-ordinated and linked to each other. The more interdependent they become, the more they learn about each other. As the individuals from the two firms work together, they gain the same knowledge and a common language is developed, as are common routines.

The description above has illustrated Powerpac's most important business relationship. But there are also other relationships that are vital for development in the firm; one instance is that of Powerpac's relationship with Lux, a producer of windows. As indicated in the presentation of the company above, Lux is Powerpac's largest customer and, in terms of the sales volume, it accounts for more than three times as much business as Playo. It is also one of the oldest customers – the relationship with Powerpac started in the mid 1950s. Powerpac is Lux's main supplier, accounting for 85–90 per cent of its packaging material. If there are any disturbances in the deliveries from Powerpac to Lux, Powerpac puts all its extra resources in to solve the problem. There have not been any major disturbances during the four decades of the relationship and it has continuously grown stronger and deeper. People from the two companies visit one another regularly and each of the companies has adapted itself extensively to the other's product and production technology. Both companies have also modified their organizational structure and how they conduct their business.

For Lux, the most important about the packaging material from Powerpac is that it has to be strong. Originally it was a rather simple, but strong cardboard in which three frames were packed. Over the years, the packaging has developed. Major improvements occurred, for instance, when Lux went from selling window-frames to complete windows with panes. Nowadays each box contains all the material required for putting together the window, with the instructions printed on the box. The box is also constructed to be opened in a certain way to protect the glass and facilitate the assembly of the window. Colour printing is requested on the packaging nowadays and the design of the boxes is more important.

Coordinating activities

The relationships with the important customers Playo and Lux also affect connected counterparts, such as suppliers and other customers. In the packaging industry, the supply of paper is essential. Two kinds are used, kraftliner and testliner. While kraftliner is mainly fabricated from new fibres, testliner is based on recycled material. Initially when Powerpac was founded in the 1950s, paper was imported from Norway. But a few years later, after the establishment of Testli, a local supplier, almost the total supply came from them. A close relationship developed between Powerpac and Testli, and the families owning the companies got to know each other very well. But from the middle of the 1960s Powerpac started to buy some kraftliner from the Swedish supplier Kraftli. In 1976 when Powerpac was bought by the Swedish MNC of which Kraftli forms a part, it was put under pressure to buy more kraftliner from Kraftli and to cut down the purchase of testliner from Testli. Nevertheless the close relationship with Testli has continued. Contact takes place on a daily basis with information sharing commonplace, despite the fact that nowadays Testli is owned by Powerpac's main competitor. This implies that old routines are difficult to change.

When a customer, for instance Playo, asks Powerpac to supply better paper quality, the requirements are passed on to the paper suppliers. As Kraftli is the largest supplier of paper, there are regular discussions between Powerpac and Kraftli. The availability of kraftliner has given Powerpac an advantage for the development of boxes for customers such as Playo, as demands from customers like this have affected the development of the paper quality. The paper quality has developed from simple brown paper during the 1970s, to white-top, a quality developed in the middle of the 1980s, which gives a much better printing quality.

Customer demand has also affected other suppliers, such as the suppliers of print. However, what is most important for Powerpac's ability to produce boxes adapted to each customer's specific requirements is the relationship with a producer of machines, Rolls. It has been possible to fulfil the customers' demands for better print quality and to satisfy the need for custom-made sizes thanks to the development of machines that Rolls has made in co-operation with Powerpac. Powerpac was the first to get the machine prototypes on which the new products were developed at the same time as the machines have been improved.

By coordinating activities across relationships, for example from suppliers to customers, learning takes place and routines are developed.

Generalising to other relationships

As the Lux and Playo examples show, Powerpac produces two very different kinds of packaging for these two important customers. While it is the print that is most important for Playo, "It has to be perfect", for Lux it is the strength of the packaging that is the determining factor. That different customers put different demands on the company is regarded as an advantage. Powerpac's production units have their own specialities and their own development staff, and these units compete with each other. When a customer asks for a new product, the units compete to come up with the most innovative solution. This internal competition within Powerpac is regarded by the managing director of Powerpac as being very important for learning and for competence development as developments in one unit of the company can be transferred to, and used by, other parts of the company. For example, the knowledge about design and printing that Powerpac has acquired from its relationship with Playo can be used in developing boxes for Lux.

As far as Powerpac's other customers are concerned, they can frequently benefit from solutions developed in the relationship between Powerpac and Playo. For Soap, a well-known multinational producer of detergent, tooth-paste, soap, etc., the development of pre-print liner has resulted in an improvement of their detergent-containing boxes. The relationship with Soap has also led to improvements that can be of use in the relationship with Playo. In the beginning of the 1990s, Powerpac and Soap co-operated intensely to develop more compact packaging for detergent, and they succeeded. Some time later, another European supplier managed to develop a similar product and sell it at a considerably lower price; this supplier now serves most detergent producers in Europe. However, the knowledge about producing compact packaging is of use for Powerpac in its ongoing development of smaller packaging with Playo.

The knowledge developed in Powerpac also spreads to customers in other countries. Although the packaging business is said to be local, Powerpac has customers abroad, mainly in Germany and Sweden. For instance, a multinational welding company based in Sweden wanted an especially strong box, that was also easy to lift, and as Powerpac's Swedish sister company could not develop this product, a box with a special lifting device was developed at Powerpac. The knowledge acquired in developing and producing boxes for Lux was of great use in this process.

In the description above it has been illustrated how learning takes place in three types of activities. It is most important to stress that learning takes place primarily when individuals perform activities in business relationships. But learning also takes place when these activities are coordinated with activities performed in connected relationships and when generalising, i.e., when routines are being developed to apply the experience in other relationships.

So far this paper has described how a subsidiary learns in its daily business activities with specific counterparts. In the following section, how knowledge is transferred to the rest of the organization will be discussed.

Knowledge diffusion within the MNC

A characteristic of learning organizations is that they not only learn from their own experience, but also from the experiences of others. They tend to put an emphasis on the rapid diffusion of knowledge throughout the whole firm. As Garvin put it (1993, p. 87): "Ideas carry maximum impact when they are shared broadly rather than held in a few hands." Hence, in order to become a learning organization, it is necessary to stimulate the exchange of ideas by opening up boundaries. "Boundarylessness" between the different units/ subsidiaries within a firm, as well as between the firm and its important counterparts, such as its customers and suppliers, ensures a fresh flow of ideas and the chance to consider alternative perspectives. The critical source of competitive advantage is knowledge integration rather than knowledge itself (Grant, 1996). A crucial task for the firm is to integrate the individuals' specialised knowledge, i.e., to encourage the organizational forms that enable learning and the exploitation of this knowledge within the firm as a whole.

Cecilia Pahlberg

Powerpac is part of an MNC consisting of a number of packaging subsidiaries spread around Europe. For the sister-units within the group, Powerpac seems to be of great importance, as it is usually the first to develop new technologies. Within the MNC of which Powerpac forms a part, the transfer of knowledge has been organized in such a way that the influence from headquarters is restricted. On the initiative of the managing director of Powerpac, an informal forum, Pacbox, was established. The idea emanated from his earlier experiences of the value of knowledge and information sharing. When the managing director was employed at Powerpac in the 1950s he got involved in an interest organization, Eurobox, comprised of packaging companies around Europe. This organization consisted of only one company from each country and Powerpac became its country's member. The purpose was to meet several times each year to exchange information and discuss R&D improvements. The international co-operation concerning product development which the company had through its membership in Eurobox stopped when Powerpac became part of the Swedish MNC as it was not the Swedish member of the organization. However, Eurobox inspired the creation of Pacbox, a similar informal organization within the group.

The purpose of Pacbox is to ensure that people meet and to facilitate the sharing of information. Through Pacbox, all subsidiaries get access to all developments and improvements within the other subsidiaries. The managing directors, the people responsible for production and development and those in charge of marketing and sales in the subsidiaries meet a couple of times each year. According to those involved, these regular meetings held on an informal basis have resulted in a continuous transfer of ideas. The co-operation within Pacbox has also resulted in common projects when people from sister-companies in co-operation develop new products. All subsidiaries in the MNC participate in Pacbox, and the chairmanship rotates, with each company taking responsibility for a year. This is important in order to delimit tendencies for the "not-invented-here syndrome" which is more likely to arise when one or just a few companies dominate, leaving the others feeling peripheral.

What is most striking is that the initiative comes from the subsidiaries and that the meetings usually take place without the participation of headquarters. The managing directors from Stockholm are only invited to attend for a limited time in order to present what happens at HQ level or to discuss a specific issue. Subsidiary managers within the group are of the opinion that these subsidiary discussions are of great value and a prerequisite to open unprejudiced discussions. As one of the subsidiary managers said, "We could easily do without HQ, but not without meeting the other subsidiaries."

However, during the last few years, after a merger with another large company and with new top management, the situation has changed and headquarters has taken control over Pacbox. It organizes the meetings and is in charge of them, which, according to people interviewed in the company, has resulted in a decrease in the information exchange.

Concluding discussion

Organizational learning is often considered to be the key to success. Among researchers, as well as practitioners and consultants, it is increasingly argued that organizational learning may be "the only sustainable competitive advantage" (Stata, 1989, p. 64, Root, 1994). In the introduction it was noted that there are different opinions about what learning is. Some stress the outcomes of learning, while others focus on the process itself. Mark Dodgson (1993) encompasses both and describes learning as "the ways firms build, supplement and organize knowledge and routines around their activities and within their cultures, and adapt and develop organizational efficiency by improving the use of the broad skills of their workforces" (p. 377). Furthermore, he points out that a main characteristic of a "learning organization" is that it facilitates learning within the organization and extends the learning culture to include important actors such as customers and suppliers (see also Pedler *et al.*, 1989).

In the case above it has been illustrated that, to a large extent, learning emanates from activities performed within specific relationships illustrated here by the relationships between Powerpac and the two customers Playo and Lux. When people from the companies involved meet and perform activities, learning takes place. The importance of specific customers is stressed by the people interviewed in the company, or as the managing director of Powerpac put it: "We live on them." The relationships are strengthened over time as the parties adapt their way of doing business to one another and routines develop.

But this case also illustrates that activities in one relationship are coordinated with activities in connected relationships. When the customers put demand on a subsidiary to develop new products, the subsidiary in its turn puts pressure on the paper suppliers, suppliers of print, the machine suppliers, etc. Hence, the firm develops routines for coordinating activities. Finally, the experiences obtained can also be generalised to, and used in, other relationships.

In MNCs, a number of learning processes are going on as each individual and subsidiary has its own knowledge base and learning capabilities. In such organizations, it is important to have a structure that encourages learning, which takes into consideration how firms may benefit from diversity and heterogeneity. As stressed by Huber, (1991), firms often do not know what they know. With the exception of systems that store "hard" information, firms usually have much weaker systems for finding where certain information exists within the organization. Thus different units with potentially synergistic information are often not aware of its coexistence. This mean that structures that facilitate information sharing are important to enable subsidiaries possessing information and subsidiaries needing this information to find one another quickly. Combining information from different subsidiaries not only leads to new information, but often also to new understanding.

As has been shown above, this development need not be centrally planned and managed. The view of the subsidiaries as integrators of knowledge provides a different perspective on the functions of organization structure. As noted by Grant (1996, p. 384): "The trend towards 'empowerment' takes account of the nature of knowledge acquisition and storage in firms: … top-down decision making must be a highly inefficient means of knowledge integration. The task is to devise decision processes that permit integration of the specialized knowledge held throughout the organization – not just in the boardroom, but on the shop floor as well." Empowerment of frontline managers, i.e., when frontline management's role is transformed from being that of an implementer to that of an initiator, creates an environment which encourages learning, co-operation and initiatives. (Garvin, 1993, p. 87). In an empowered organization, the headquarter's role is to provide a context in which this can happen. As the case shows, a worldwide learning process, which encourages the units within an MNC to share information, might benefit from the absence of participation and control from headquarters.

References

BARTLETT, C., and GHOSHAL, S., 1989, Managing Across Borders: The Transnational Solution. Harvard, Boston, MA.

BIRKINSHAW, J., 1997, Entrepreneurship in Multinational Corporations: The Characteristics of Subsidiary Initiatives *Strategic Management Journal*, Vol. 18, No. 3, pp. 207–229.

DODGSON, M., 1993, Organizational Learning: A Review of Some Literatures *Organization Studies*, Vol. 14, No. 3, pp. 375–394.

DUNCAN, R., and WEISS, A., 1979, Organizational Learning: Implications for organizational design. In Staw, B., and Cummings, L.L., (eds), *Research in organizational behaviour*, pp. 75–132, JAI, Greenwich, CT.

ERIKSSON, K., HOHENTHAL, J., and JOHANSON, J., 1998, A Model of Learning in International Business Networks, *WZB Yearbook 1998 Learning*, Berlin: Wissenschaftszentrum.

FORSGREN, M., and JOHANSON, J., 1992, *Managing Networks in International Business*, Philadelphia: Gordon and Breach Science Publishers S.A.

FURU, P., 1997, Conceptualizing Centers of Excellence in Multinational Corporations. Paper presented at AIB Conference in Mexico.

GARVIN, D., 1993, Building a Learning Organization *Harvard Business Review*, July-August, pp. 78–91.

GHOSHAL, S., and BARTLETT, C., 1990, The Multinational Corporation as an Interorganizational Network *Academy of Management Review*, Vol. 15, No. 4, pp. 603–625.

GRANT, R., 1996, Prospering in Dynamically-competitive Environments: Organizational Capability as Knowledge Integration *Organization Science*, Vol. 7, No. 4, pp. 375–387.

GUPTA, A., and GOVINDARAJAN, V., 1991, Knowledge Flows and the Structure of Control within Multinational Corporations *Academy of Management Review*, Vol. 16, No. 4, pp. 768–792.

GUPTA, A., and GOVINDARAJAN, V., 1994, Organizing for Knowledge Flows within MNCs *International Business Review*, Vol. 3, No. 4, pp. 443–457.

HÅKANSSON, H., 1987, *Industrial Technological Development – A Network Approach*, London: Croom Helm.

HIPPEL, E. VON, 1988, *The Sources of Innovation*, Oxford: Oxford University Press.

HUBER, G., 1991, Organizational Learning: The Contributing Processes and the Literatures *Organization Science*, Vol. 2, No. 1 pp. 88–115.

KANTER, R. M., 1982, The Middle Manager as Innovator *Harvard Business Review*, pp. 95–105.

MILLER, D., 1983, The correlates of entrepreneurship in three types of firms *Management Science*, Vol. 29, pp. 770–791.

MINER, A., and MEZIAS, S., 1996, Ugly Duckling No More. Pasts and Futures of Organizational Learning Research *Organization Science*, Vol. 7, No. 1, pp. 88–99.

PEDLER, M., BOYDELL, T., and BURGOYNE, J., 1989, Towards the Learning Company *Management Education and Development*, Vol. 20, No. 1, pp. 1–8.

ROOT, H. P., 1994, MSI: A resource for the Learning Organization *Marketing Science Institute Review*, Spring.

STATA, R., 1989, Organizational Learning: The Key to Management Innovation *Sloan Management Review*, pp. 63–44.

WEICK, K., 1991, The Nontraditional Quality of Organizational Learning *Organization Science*, Vol. 2, No. 1, pp. 116–124.

Part III Company Network Learning

CHAPTER 11

Relationship Configuration and Competence Development in MNC Subsidiaries

MARIA ANDERSSON, ULF HOLM and CHRISTINE HOLMSTRÖM

Introduction

One notion stressed in research on the multinational corporation (MNC) lately is the importance of generating and utilizing knowledge[1] and resources from subsidiaries located in different parts of the world as a strategy for gaining competitive advantage (Gupta and Govindarajan 1994; Bartlett and Ghoshal 1989). Another notion is that knowledge development among MNC members may be more than just an intra-corporate matter although it has been shown that the corporation is essential for innovation and for the control and distribution of corporate knowledge (Kogut and Zander 1993). Thus, the knowledge that subsidiaries create is to some extent externally driven and as Ghoshal and Bartlett point out (1990), the dependence on resources in an external exchange network can explain the distribution of the MNC's internal resources. As the distribution of corporate resources affects, for example, changes in technology and business conduct, those subsidiaries that mediate the impact of external resource dependencies are of crucial importance for the long-term development of an MNC (Forsgren, Holm and Thilenius 1997).

---

[1]In the literature concepts like knowledge and competence have somewhat various meanings. In this paper, these words are used as synonyms.

In the context of the MNC, it has also been argued that the role and the importance of the market network can be related to the internationalization process of the firm (Forsgren 1989). Usually, once a green-field foreign subsidiary is created, it will at first be highly dependent on corporate resources and on headquarters' decisions about the distribution of resources. However, in becoming established, the typical foreign subsidiary will engage in local business activities with actors external to the MNC and it will gradually become more dependent on the resources exchanged within the market network. The adoption of such an inter-organizational perspective on the MNC has arisen through the supposition that the business settings of subsidiaries affect their operational competence (Amit and Schoemaker 1993, Grant 1991, Mahoney and Pandian 1992, Peteraf 1993, Teece, Pisano and Shuen 1990).

Therefore, in studying competence development, and in line with Ghoshal and Bartlett (1990) and Grant (1996), we have reason to include market relationships along with the corporate ones. Recent empirical findings, on subsidiary relationships and their importance for technological and business development, support such an approach as 80 per cent of the most important relationships have been identified as being external to the MNC (Andersson and Pahlberg 1997, Thilenius 1997). There is consequently a need to systematically study the impact of the market network.

It should be noted that the inter-organizational approach to the MNC has meant a shift in focus when studying MNC development. Firstly because it puts the subsidiary in focus to a larger extent, because it is a crucial link for the accumulation of knowledge generated from business activities in various market networks. Secondly, subsidiaries have specific corporate roles and competencies and contribute to MNC development in different ways. The role that market networks play in this matter is a complex one since MNCs usually consist of many subsidiaries, all with their specific network structures and each with the potential to have an impact on the development of competencies in the other subsidiaries. Thirdly, as competence development includes both market and corporate relationships, it is not the legal boundary of the corporation that sets the limit for the analytical focus. Rather, the relationships with those counterparts perceived to be important by the subsidiary, independent of whether they are corporate or market, determine the context of the analysis (Snehota 1990, Håkansson and Snehota 1989).

An MNC subsidiary will have a configuration of important relationships that is at least partly external to the MNC. Still, received theory deals to a considerable extent with subsidiary competence development as a matter of MNC distribution and the integration of knowledge, or as a matter of development within the local subsidiary organization (Etemad and Dulude 1986). Although the market network of the MNC is generally acknowledged, there has been limited research examining and comparing its impact on subsidiary competence development and comparing it with the impact of the corporate network of the multinational.

In this paper, the term *subsidiary relationship configuration* refers to the set of relationships that the subsidiary depends on for its competence development, e.g., corporate (MNC) relationships, market relationships, or both. This is illustrated in Figure 11.1. Moreover, the study focuses on critical relationships with subsidiary suppliers, customers and R&D units. Those of the relationships that belong to the same formal organizational structure as the subsidiary, i.e., to the MNC, are labeled *corporate relationships*, whilst *market relationships* is the label for subsidiary relationships beyond the formal boundaries of the MNC.

FIGURE 11.1

The Subsidiary Network – Market and Corporate Relationships

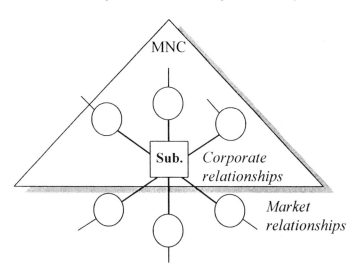

The purpose of this paper is to study the impact of the corporate and market relationship on subsidiary competence in three activities: the development of products and processes, the production of goods and services, and marketing and sales. In the analysis the level of subsidiary competence among four groups of subsidiaries will be tested. These groups are based on the existence of critical market and corporate relationships. That is, the subsidiaries are divided up according to whether they have no critical relationships, only critical corporate relationships, only critical market relationships or, finally, critical relationships with both market and corporate actors.

In the following section we argue for the relevance of studying the impact of long-term relationships on competence development. We then discuss the research method and present empirical data on a total sample of 3346 subsidiary relationships and 577 subsidiaries in Denmark, Finland, Norway and Sweden. The difference that the impact of relationships has on competence development in the four groups of subsidiaries was tested through an ANOVA analysis. Finally, the results are presented and issues for further research are proposed.

Competence and competence development

The following section discusses the meaning of competence and the issues related to learning and the transfer of competence between business actors. The subsequent section then examines the importance of long-term interaction and the creation of critical relationships as explanations of competence development.

Competence is sometimes described as a matter of "being able to do something well" (Oxford Advanced Learner's Dictionary, 1995) and being competent as a matter of "having the ability, authority, skills, knowledge, etc. to do what is needed" (ibid.). But for an organization to have ability to do what is needed, specific competencies must be stored and shared amongst and between organizational members, i.e., there is a learning process based on the utilization of knowledge and on informing individuals. Thus, competence development is a matter of organizational learning (Snow and Hrebiniak 1980). But to reach an integrated organizational competence, knowledge must be distributed throughout the organization and become embodied into products, services and systems (Nonaka and Takeuchi 1998). An underlying dilemma is that knowledge varies in the degree to which it is articulated, and is more or less tacit. This type of knowledge cannot easily be communicated using words or numbers, and thus it is difficult to codify (Winter 1987; Polanyi 1962). To be integrated in the organization social relations are required, i.e., there must be personal contact (Grant 1996).

In a business context competence is connected with the ability to generate and take advantage of business opportunities (Patel and Pavitt 1998). This means that the competence of a company is not separated from its business environment. In fact, the ability of organizations to learn and generate higher levels of knowledge depends on environmental interaction (Cyert and March 1963, Forsgren, Johanson and Sharma 2000, Nelson and Winter 1982). But the learning problem varies depending on the activity being investigated. Compared to activities that involve the exchange of capital, raw material, patents and licences, activities such as the development of technology, of production processes and of marketing know-how need a closer relationship based on interaction for learning to occur. The reason for this is that these latter activities are relatively tacit (Makhija and Ganesh 1997). Therefore developing a high level of competence in these areas, through contact with counterparts, is dependent upon the kind of interaction and relationships that the parties have. This indicates, for example, that the development of a new product is the result of an adaptation to and the successful integration of the knowledge contained within the relationships by the parties/individuals involved (Clark and Fujimoto 1991).

It should be stressed that our view on the competence development of an MNC subsidiary, as far as production, development and marketing is concerned, is not a matter of having tacit knowledge codified and transferred in a straightforward manner between counterparts. Rather, the competence of subsidiaries in these business activities is developed through interaction with other parties; a process of learning-by-interacting, implying incremental learning (Uzzi 1997; Grant 1996). We can therefore expect that a specific company's ability to develop new competence through contacts with another company depends on how the parties interact and what kind of competence is involved. This implies that the building of relationships is an important means with regard to competence development within MNC subsidiaries.

The impact of relationships on subsidiary competence

Arguing that competence development is a learning-by-interacting process (Grant 1996), some business relationships will impact the institutionalization of knowledge within the subsidiary organization and can be regarded as critical for its level of competence. By critical, we mean the decisive impact that a relationship has on a subsidiary's competence development based on the degree of dependence on that relationship (Pfeffer and Salancik 1978). Therefore, the configuration that comprises a subsidiary's most critical relationships exhibits an important analytical basis for the understanding of subsidiary competence development. But before we elaborate on the possible configurations of subsidiary relationships, we will discuss further the emergence of critical relationships in a network.

As the theoretical development on business networks has identified there is a distinction between exchange relations and close relationships. An exchange relation is characterized by mere buying and selling activities, i.e., one-shot deals, focusing on costs (Johanson and Mattsson 1994). It lacks social content, and thus it does not involve any deeper commitments from the parties involved (Powell 1990). Other characterizing features of exchange relations are infrequent interaction and irregular, scarce information flows. In comparison, when the interactions are intense, it means that a business unit often involves larger volumes of business exchange and several common technological issues to solve with customers, suppliers and other counterparts. This requires deep knowledge about the counterpart organization and thus, when companies have common interests close co-operative relationships are likely to emerge.

Such relationships are also associated with mutual dependence between the partners, which perpetuates the co-operation (Hallén 1982; Håkansson 1987; Cowley 1988; Perrone 1989). Hence, we can expect that relationships that are well established and critical lead to an intention to remain in the relationship and give the parties involved confidence of future exchange and of common development projects (Anderson and Weitz 1989; Morgan and Hunt 1994; Doney and Cannon 1997). Powell (1987) argues that established relationships promote learning as they are a means of tapping into sources of know-how located outside the boundaries of the company. In comparison to ordinary exchange relations, such relationships convey advantages as they serve as a means for the exchange of tacit information and proprietary know-how, and thus promote learning (Uzzi 1997). It should be stressed, though, that not all exchange relations evolve into close relationships (Ford 1980). In fact, close and critical relationships are relatively few in number since limitations on the organizational capacity, such as limitations on the time and the manpower, restrict a business actor from engaging deeply in more than a limited set of business relationships (Ford 1990 Håkansson 1989 Uzzi 1997).

Assuming that an MNC subsidiary engages deeply in a limited set of critical relationships, we propose that such relationships have a positive effect on the subsidiary competence in development, production and marketing activities. The issue here is to identify different subsidiary relationship configurations consisting of corporate and/or market relationships, and to analyse their impact on subsidiary competence development. We anticipate that subsidiaries with critical corporate and market relationships will have higher levels of competence than subsidiaries without such relationships. In the next section the impact of different relationship configurations will be discussed.

Subsidiary relationship configuration

Against this background, we can picture different relationship configurations for an MNC subsidiary. In Figure 11.2 we employ four groups of subsidiaries according to how they are configured in terms of corporate and/or market relationships. In the first group, no relationships have a critical impact on subsidiary competence. The second group consists solely of corporate relationships with a critical impact, and the third is a situation when only market relationships have a critical impact. In the fourth group, both corporate and market relationships have a critical impact on subsidiary competence.

FIGURE 11.2
Corporate and market subsidiary relationships with a critical impact on competence development

One
or
more

Market relationships
with a critical impact on
subsidiary competence

	3	4
	1	2

None

None   One or more

Corporate relationships
with a critical impact on
subsidiary competence

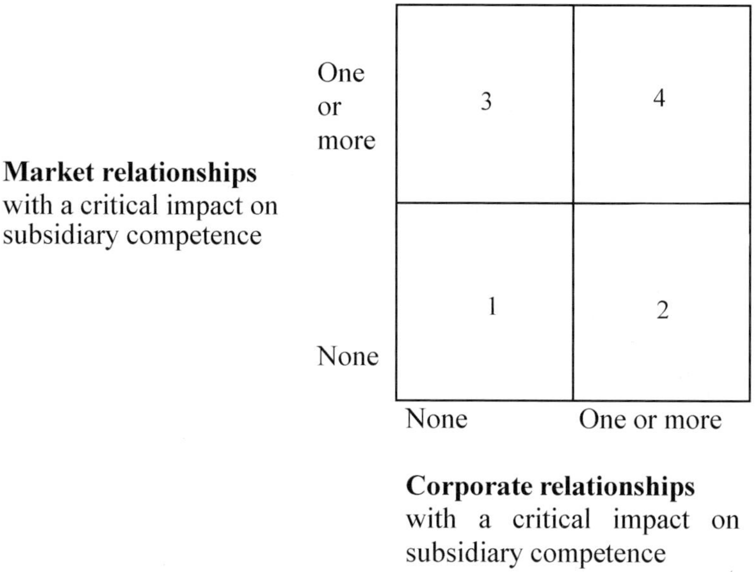

In the lower left hand corner (1) we have subsidiaries that have no specific business relationships with a critical impact on competence development. The subsidiary relies mainly upon exchange relations or arm's-length deals (Uzzi 1997) characterized by non-repeated interaction. This kind of subsidiary is by no means isolated from the environment, but it has no close interactions in the sense that no specific relationships within the MNC or with counterparts in the external market have a critical impact on its competence development.

In the lower right hand corner (2) we have the subsidiaries embedded in corporate relationships with a critical impact on competence development. Interacting within the MNC means that the subsidiary not only relates to its counterparts through autonomous and personal interaction, but it also involves MNC directions manifested in, e.g., common MNC operating manuals and organizational routines (Grant 1996; Nelson and Winter 1982). According to Hedlund (1986), it can be expected that the creation of common norms, values and standards will facilitate knowledge integration within the MNC, and this knowledge will be conveyed through the relationships between organizational MNC members (Kogut and Zander 1993).

However, the impact of the corporate system on the competence of a specific subsidiary can be questioned because (the use of) shared norms, values and standards can work in both ways. These mechanisms may create a system of effective and regular flows of information (as argued above), but there is a risk that they will impede subsidiary learning because they may be inadequate for the specific business activities of the single subsidiary (Grant 1996). Still, given the risk of internal inflexibility, the corporation may be effective and, according to Grant (ibid.) intra-firm relationships should be superior in the integration of knowledge between organizational units. Thus, the corporation serves the purpose of knowledge development.

In the upper left hand corner (3) we find subsidiaries that benefit from interacting with one or several counterparts on the external market. In the typology introduced by Andersson and Forsgren (1995), this configuration of relationships corresponds to "the external subsidiary", meaning that the MNC and the subsidiary primarily have administrative or financial links. Empirical studies have shown that market relationships are of great importance for technological development (Andersson and Pahlberg 1997). As Grant (1996) argues, the fundamental advantage of an external set of knowledge linkages in comparison with having corporate relationships, is the flexibility and autonomy in dealing with a wider set of possible counterparts.

Finally, in the upper right hand corner (4) subsidiaries develop their competence on the basis of critical impact from both corporate and market relationships. In this situation the subsidiary organization is under impact of a variety of relationships. For instance, a corporate supplier and an external customer may both affect the competence of a subsidiary. Given that the subsidiary can utilize both corporate and market relationships we can expect it to develop a higher level of competence. However, there may be a problem of consistency between belonging to a corporate context with certain pressures concerning how to develop products and processes in relation to the demands of the market relationship context (Forsgren 1997).

Research method

To analyse the impact of relationships on subsidiary competence we needed to gather information that is comparable from subsidiary to subsidiary. This meant that we had to decide: (i) what competence to measure; and (ii) what subsidiaries and subsidiary relationships to include.

The sample

This specific study of subsidiary competence concerns three activities: production (of goods or services), development (of products and processes) and marketing and sales. These activities were chosen as they are usually regarded as contributing to the company's long-term development (von Hippel 1988) and as they are related to the building of relationships with counterparts in the subsidiary environment (Makhija and Ganesh 1997). As it has been decided to concentrate on those subsidiaries with competence in production, development and marketing, it follows that the sample includes so-called fully fledged subsidiaries, i.e., units that perform several operational functions (Etemad and Dulude 1986). By employing these criteria we exclude subsidiaries that only perform single activities, such as sales or basic research.

One fundamental assumption is that identifying a limited set of critical relationships will provide an adequate set for the analysis of subsidiary competence. We include three relationships, i.e., those with the most critical supplier, customer and R&D unit within the market and the corporation respectively. Thus, altogether, the criticality of six subsidiary relationships is investigated. The reason for choosing these relationships is that business relationships with suppliers, customers and R&D units have been shown to be important for the technological and business development of a business unit (Håkansson 1987, 1989; von Hippel 1988; Clark and Fujimoto, 1991). The relationship with corporate HQ is not included because by definition the subsidiary network involves lateral relationships to counterparts of importance for the business development (Ford 1990). Although potentially important, the hierarchical control relationship with HQ is not primarily such a relationship.

Data collection and measurement

The data collection was done by survey, i.e., questionnaires were sent out to subsidiaries in Denmark, Norway, Finland and Sweden by participating researchers in the Nordic countries.[2]

The questionnaire was directed to and answered by subsidiary managers as they, compared to MNC top managers, can be assumed to be more familiar with the business relationships that drive the competence development in the subsidiary, 80 per cent of the respondents of the questionnaire were subsidiary executive officers. The remaining percentage consisted of financial managers, marketing managers and controllers.

When measuring the competence of the subsidiary in production, development and marketing, the respondent was asked to evaluate his/her

---

[2]This study is done within the frame of the international project "Centre of Excellence" which focuses on MNC subsidiaries with foreign HQs. The databases Ekonomisk Litteratur (Sweden), Dun & Bradstreet (Finland and Norway) and Greens (Denmark) were used to identify such subsidiaries.

company's competence on a 7-point Likert scale, ranging from weak to very strong. In a corresponding way, the respondent was asked to indicate the impact of business relationships on the development of subsidiary competence on a 7-point scale, ranging from no impact to a very decisive impact. To reduce the level of missing values in the survey, contact was remade with several respondents and completion of the questionnaires was made.

The data gathering procedure resulted in a low level of missing values, 3.4 per cent for the measured impact of all market relationships and 2.5 per cent for all corporate relationships. The level of missing values on subsidiary competence in the three activities is 1.7 per cent. The number of subsidiaries sampled in the four countries investigated was 577. Together, they have provided information on a total of 3346 business relations to customers, suppliers and R&D units, of which 1672 were corporate relationships and 1674 were market relationships. All kinds of subsidiary companies within the service and manufacturing industries are represented. The size of the subsidiaries in this final sample ranges from 3 to 9.300 employees with an average of 377 and an annual turnover ranging from 0.1 to 2.884 million USD, with an average of 94.

An important issue for the analysis of subsidiary competence is the identification of those subsidiary relationships that are critical for the development of competence in the subsidiaries. Therefore we selected the market and corporate relationships with a perceived impact level of 6 and 7 on the seven-point scale. From that basis we then distributed the subsidiaries according to the structural configuration illustrated in Figure 11.2. Thus, those subsidiaries that fall into the lower left hand corner are regarded as having no relationships with a decisive impact on the development of competence in their subsidiary. Those having decisive relationships are distributed among the three other groups depending on whether relationships are with the market, within the MNC or both.

It should be noticed that the measures of the degree to which the relationship has an impact on the subsidiary competence are based on the perception of the subsidiary managers (Zucker 1991). Thus, the characteristics of specific relationships do not have any objective attributes and as argued by, for example, Kelly and Thibaut (1978), perception is the reality.

Empirical basis

The data analysis is carried out as follows. First we present a descriptive analysis of the existence of market and corporate relationships with decisive impact on subsidiary competence (Table 11.1). We also illustrate the distribution of subsidiary competence in the three activities studied. The analysis then proceeds by performing an ANOVA analysis: examining the subsidiary level of competence within the four subsidiary relationship groups.

TABLE 11.1

Share of market and corporate relationships having a decisive impact on subsidiary competence development

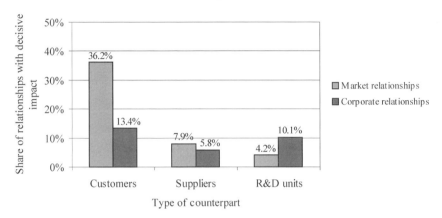

Table 11.1 shows that the existence of market and corporate relationships with a decisive impact on subsidiary competence differs somewhat. While 36.2 per cent of all the market customers have a decisive impact, the corresponding figure for corporate customers is only 13.4 per cent. The supplier values are lower, but show a similar number, i.e., 7.9 and 5.8 per cent for market and corporate relationships respectively. Decisive impact from relationships with R&D units is more frequent among the subsidiary's corporate relationships (10.1 per cent) than among market relationships (4.2 per cent).This means that for subsidiaries with only market relationships it is very common that customers are the main drivers of competence development, while there is a more even distribution of counterparts having a decisive impact among subsidiaries with only corporate relationships.

This also means that it is more common for subsidiaries to have critical market relationships than corporate relationships, at least when using values 6 and 7 as indicators of decisive impact. Figure 11.3 reveals that of the 577 subsidiaries investigated, 146 (25.3%) belonged to the category that had one or several market relationships, 43 (7.5%) belong to the category with one or several corporate relationships, 87 (15.1%) have both market and corporate relationships and as many as 301 (52.2%) have neither market nor corporate relationships.

Table 11.2 portrays the distribution of subsidiary competence in production, development and marketing. The values show that the majority lie in the upper range of the seven-point scale. The average competence in production is 5.87, slightly higher than both development (5.21) and marketing (5.69). The majority of the subsidiaries, 69.8 per cent, claim that they have a very strong competence in production (values 6 and 7) while the corresponding figures are 47.5 per cent for development and 63.4 per cent for marketing.

Analysis of relationship impact on subsidiary competence development

For the empirical analysis of subsidiary competence, we employ the subsidiary relationship configuration outlined in Figure 11.3. We analyze subsidiary competence in development, production and marketing with regard to the four groups of subsidiaries. Group 1 consists of subsidiaries with no critical relationships. Groups 2 and 3 are made up of subsidiaries with critical corporate and market relationships respectively. Finally, Group 4 consists of subsidiaries with both critical corporate and market relationships.

The analysis was done using an ANOVA and a least significant test (LSD) to see whether there are significant differences between the groups. ANOVA is commonly used to determine whether deducted groups are significantly separated from each other (Hassmén and Koivula 1996). The aim is to determine whether variation in subsidiary relationship

FIGURE 11.3

Number of subsidiaries in each of the four configurations of corporate and market relationships that have a decisive impact on the development of competence

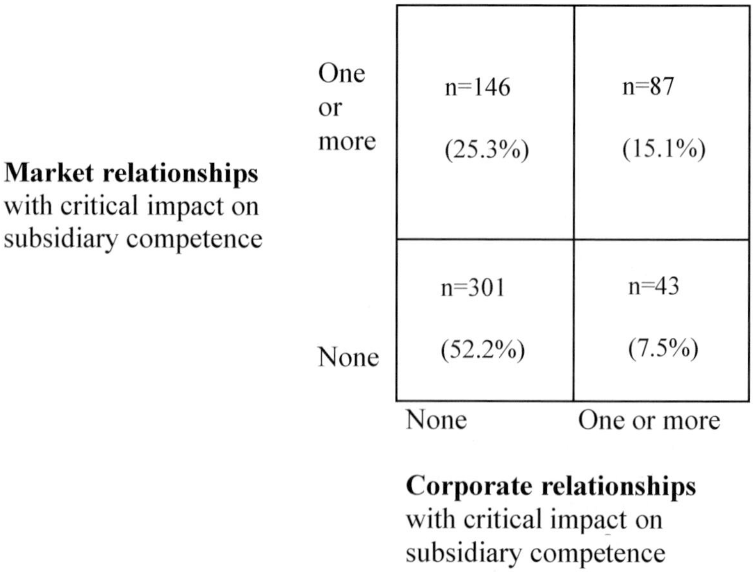

Market relationships with critical impact on subsidiary competence

One or more: n=146 (25.3%), n=87 (15.1%)
None: n=301 (52.2%), n=43 (7.5%)

None | One or more

Corporate relationships with critical impact on subsidiary competence

TABLE 11.2

Distribution of perceived competence in MNC subsidiary activities

	Subsidiary activity		
Degree of perceived subsidiary competence	Production of goods and services n (%)	Development of products and processes n (%)	Marketing and sales activities n (%)
1	3 (0.5)	2 (0.4)	2 (0.4)
2	3 (0.5)	8 (1.4)	4 (0.7)
3	12 (2.1)	66 (11.6)	18 (3.2)
4	40 (7.1)	87 (15.3)	60 (10.5)
5	113 (20.0)	135 (23.8)	123 (21.6)
6	215 (38.0)	166 (29.3)	215 (37.8)
7	180 (31.8)	103 (18.2)	147 (25.8)
Total Average	566 (100.0) 5.87	567 (100.0) 5.21	569 (100.0) 5.69

configuration can explain differences in the level of competence in subsidiaries. The results are presented in Table 11.3.

First we can state that the significant p-values ($p < .05$) provide evidence that there are significant differences between the four different groups of subsidiaries with regard to the subsidiary competence in all three activities tested. The result therefore confirms our general proposition that development of competence can be explained by the existence of a limited set of critical relationships in the subsidiary environment.

The results show that group (2), the group of subsidiaries with only corporate relationships, has a significantly higher competence in all three activities than has group (1), with no critical relationships. This suggests that when subsidiaries are involved in corporate relationships, they tend to develop strong competence. The group of subsidiaries with only market relationships, (3), shows a significantly higher competence in two activities, development and production, than the group without critical relationships (1). Finally, the subsidiaries that combine market and corporate relationships (4) demonstrate higher competence in production and marketing than the group without critical relationships.

Thus, as portrayed in Figures 11.4–6 and based on the analysis shown in Figure 11.3, the relevance of critical relationships being corporate, market, or both, depends on the activity in focus.

Subsidiary competence at production is not sensitive to corporate boundaries as all three groups of subsidiaries (2, 3 and 4) have significantly higher competence levels than subsidiaries with no critical relationships. It

TABLE 11.3
Business relationship configuration and impact on subsidiary competence

Subsidiary Competence	1 No relationships	2 Corporate relationships	3 Market relationships	4 Corporate and market relationships	F-Statistic	P-Value	Groups' sign. different at p<.05 (ANOVA)[1]
Development	5.07	5.65	5.35	5.27	3.34	0.019	$2^{**}, 3^{*} > 1$
Production	5.71	6.14	5.97	6.11	4.76	0.003	$2^{*}, 3^{*}, 4^{**} > 1$
Marketing	5.56	5.95	5.77	5.89	3.29	0.020	$2^{*}, 4^{*} > 1$

1: **, * denote significance at the 1% and 5% levels respectively.

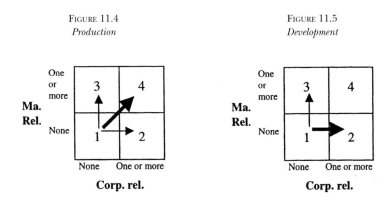

FIGURE 11.4
Production

FIGURE 11.5
Development

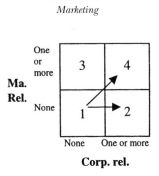

FIGURE 11.6
Marketing

seems that relationships matter independent of whether they are corporate or market, as long as they have a critical impact. This is strengthened by the fact that the combination with both corporate and market relationships increases the significance between the groups.

For development activities, the results indicate that either corporate relationships or market relationships drive subsidiary competence. But the combination of corporate and market relationships does not lead to a significantly higher competence. The corporate boundaries seem to matter in this respect. When comparing development with production, it seems that the development of competence in the subsidiary benefits from engaging in a limited set of either corporate or market relationships, while competence in production benefits from the general impact of several relationships, independent of the corporate or market context.

For the subsidiary's marketing and sales activities, corporate relationships and the combination of corporate and market relationships increase the subsidiary competence. It should be noted that, although by definition the activity is highly related to external market actors, it is only in combination with corporate relationships that relationships with these actors matter.

Comparing Tables 11.1 and 11.3 reveals an interesting observation: Although it is more common to have market relationships with a decisive impact, i.e., market customers, this does not necessarily lead to subsidiaries with higher competence as compared with subsidiaries having only corporate relationships with decisive impact. Finally, although the analysis reveals that relationships matter, it should be pointed out that more than 50 per cent of the subsidiaries did not perceive relationships as being critical for their competence development. This has to some extent to do with the choice of criticality level of relationship impact in our analysis.

Conclusions and discussion

The conclusion of this study is that it is not only corporate relationships, but also external market relationships that matter and that must be included when analyzing the competence development of MNC subsidiaries. However, this does not mean that corporate relationships are unimportant. On the contrary, the latter group of relationships is essential for all three subsidiary activities analyzed in this study.

The second conclusion is that the impact of market and corporate relationships, separately and in combination, has to do with the specific subsidiary activity in focus. The development of products and processes seems to be more sensitive to a limited set of relationships than the other two activities studied. According to Makhija and Ganesh (1997), this activity involves a larger portion of tacitness and therefore imposes limitations on the subsidiaries' possibilities of carrying out development outside a close group of relationships. In this case, the corporate context and the market context clearly are separate, although both seem important. Production competence relates more to the whole system of relationships and responds to the general pressure from the relationship network. Somewhat surprisingly, the marketing and sales competence of a subsidiary is primarily connected to corporate relationships and/or the combination of market and corporate relationships. This may have to do with the subsidiary use of corporate expertise and experience in marketing from other MNC countries and corporate standard instructions and technical manuals used for the marketing of products.

From a managerial perspective we have reason to distinguish between managing the corporate utilization of subsidiary competence and managing the development processes leading to distinctive subsidiary competence. The complicated pattern and content of specific subsidiary relationships mean that it is questionable whether the latter process can be managed by the MNC HQ, especially in cases where market relationships are the driving forces. According to Krackhardt (1990), network knowledge is critical for control and coordination, and research on business networks has indicated that such knowledge is not easily obtained, especially for an actor which is not directly involved in the interactions of a certain network.

Another complicated procedure is managing the corporate utilization of a given subsidiary competence, manifested in the ability to develop new products or process technology. However, Kogut and Zander (1995) found that the corporation is efficient at transferring and utilizing subsidiary knowledge for the benefit of the whole MNC. But at the same time, we can expect that subsidiary competence is to a considerable extent related to conducting business with local customers and suppliers, meaning that it may be high, but context specific. Thus, the subsidiary may have strong competence that is of little use to others within the MNC, implying a lack of absorptive capacity (Szulanski 1996).

The corporate ability to use a subsidiary's competence elsewhere in the MNC is also dependent upon the MNC's ability to recognize the subsidiary's competence. The size of MNCs and the specific character of subsidiary market relationships prohibits the MNC from having unlimited knowledge of the character and the location of MNC competencies and consequently, of their possible corporate usability. Furthermore, the more that other MNC units use the subsidiary competence the more the subsidiary will control a resource on which they become dependent for their activities. To the extent that its competence is derived from external market relationships, the subsidiary constitutes the corporate link to these resources and its structural position offers the possibility to affect resource configuration decisions within the MNC (Ghoshal and Bartlett 1990, p. 91). This may not necessarily mean that the subsidiary supports the corporate integration of its competence as it may prefer to operate autonomously rather than engaging in corporate exchanges.

The analysis of the different relationship structures in this study has been restricted to identifying the difference in competence between groups of subsidiaries. An issue for future research is to more systematically test the variation of different relationships' impact on different types of subsidiary competence. One hypothesis derived from our results is that competence development in activities that include a larger portion of tacitness needs a more context specific and delimited set of relationships. The dilemma is that the more context specific such relationships are, the lower the possibility is that one can utilize the subsidiary competence in other contexts, such as elsewhere in the MNC (Forsgren 1997; Forsgren, Johanson and Sharma 1999; Holm and Pedersen 2000).

Another issue for future research is to investigate how over time interactions in relationships lead to manifestations of high competence in the development activities, production technology and marketing. For this issue we expect a network approach to be useful, meaning that the impact of specific relationships may be connected to other relationships within and outside the MNC, leading to a better understanding of how the competence of a subsidiary develops on the basis of different relationship configurations.

References

AMIT, R., and SCHOEMAKER, P. J. H., 1993, Strategic Assets and Organizational Rent *Strategic Management Journal*, Vol. 14, pp. 33–46.

ANDERSON, E., and WEITZ, B., 1989, Determinants of continuity in conventional industrial channel dyads *Marketing Science* 8 (Fall), pp. 310–23.

ANDERSSON, U., and FORSGREN, M., 1995, Using networks to determine multinational parental control in subsidiaries in Paliwoda, S.J., and Ryans, J.K., (eds) *International Marketing Reader*, London: Routledge.

ANDERSSON, U., and PAHLBERG, C., 1997, Subsidiary influence on strategic behaviour in MNCs: An empirical study *International Business Review*, 6(3): pp. 319–334.

BARTLETT, C. A., and GHOSHAL, S., 1986, Tap your subsidiaries for global reach *Harvard Business Review*, 64(4): pp. 87–94.

BARTLETT, C.A., and GHOSHAL, S., 1989, *Managing Across Borders: The Transnational Solution*, Boston: Harvard Business School Press.

CLARK, K. B., and FUJIMOTO, T., 1991, *Product Development Performance: Strategy, Organization and Management in the World Auto Industry*, Boston, Mass.: Harvard Business School Press.

COWLEY, P. R., 1988, Market structure and business performance: an evaluation of buyer/seller power in the PIMS database, *Strategic Management Journal* 9: pp. 271–278.

CYERT, R, and MARCH, J. G., 1963, *A behavioural theory of the firm*, Englewood Cliffs, NJ: Prentice Hall.

DONEY, PATRICIA M., and CANNON, JOSEPH P., 1997, An examination of the nature of Trust in Buyer-Seller Relationships, *Journal of Marketing*, Vol. 61 (April), pp. 35–51.

ETEMAD, H., and DULUDE, L. S., (eds), 1986, *Managing the multinational subsidiary – responses to Environmental changes and to Host Nation R&D policies*, London: Croom Helm.

FORD, D., 1980, The development of buyer-seller relationships in industrial markets *European Journal of Marketing*.

FORD, D. (ed.), 1990, *Understanding Business Markets: Interaction, Relationships, Networks*, London: Academic Press.

FORSGREN, M., 1989, *Managing the Internationalization Process – the Swedish Case*, London: Routledge.

FORSGREN, M., 1997, The advantage paradox, in Björkman, I., and Forsgren, M., (eds), *The nature of the international firm – Nordic contributions to international business*, Copenhagen: Handelshøjskolens forlag.

FORSGREN, M., JOHANSON, J., and SHARMA, D., 2000, Development of MNC Centres of Excellence, in Holm, U., and Pedersen, T., (eds), The Emergence and Impact of MNC Centres of Excellence: A Subsidiary Perspective, (forthcoming), London: MacMillan.

FORSGREN, M., HOLM, U., and THILENIUS, P., 1997, Network infusion in the multinational corporation, in Björkman, I., and Forsgren, M., (eds), *The nature of the international firm – Nordic contributions to international business*, Copenhagen: Handelshøjskolens forlag.

GHOSHAL, S., and BARTLETT, C. A., 1990, The multinational corporation as an interorganizational network *Academy of Management Review*, 15(4): pp. 603–25.

GRANT, R. M., 1991, The Resource-Based Theory of Competitive Advantage: Implications for Strategy Formulation, *California Management Review*, Spring.

GRANT, R. M., 1996, Prospering in Dynamically-competitive Environments: Organizational Capability as Knowledge Integration *Organization Science*, Vol. 7, No. 4 July-August.

GUPTA, A., and GOVINDARAJAN, V., 1994, Organizing for knowledge flows within MNCs *International Business Review*, 3(4): pp. 443–457.

HÅKANSSON, H., 1987, *Industrial Technological Development – A Network Approach*, London: Croom Helm.

HÅKANSSON, H., 1989, *Corporate Technological Behaviour – Co-operation and Networks*, London: Routledge.

HÅKANSSON, H, and SNEHOTA, I., 1989, No Business is an Island: The network concept of business strategy, *Scandinavian Journal of Management*, Vol. 5, No. 3. pp. 187–200.

HALLÉN, L., 1982, *International Industrial Purchasing. Channels, Interaction and Governance Structure*, Uppsala University (Reprint no. 13).

HASSMÉN, P., and KOIVULA, N., 1996, *Variansanalys*, Lund: Studentlitteratur.

HEDLUND, G., 1986, The hypermodern MNC: A heterarchy? *Human Resource Management*, 25 (1): pp. 9–25.

HIPPEL, E. VON, 1988, *Sources of Innovation*, Oxford: Oxford University Press.

HOLM, U., JOHANSON, J., and THILENIUS, P., 1995, Headquarters' Knowledge of Subsidiary Network Contexts in the Multinational Corporation, International Studies of Management and Organisation Vol. 25, Nos. 1–2, pp. 97–119. M.E. Sharpe, Inc., 1995.

HOLM, U., and PEDERSEN, T., (eds), 2000, *The Emergence and Impact of MNC Centres of Excellence: A Subsidiary Perspective*, London: MacMillan.

JOHANSON, JAN, and MATTSSON, L- G., 1994, The Market-as-Networks Tradition in Sweden in Laurent, G., and Lilien, G.L., (eds), *Research Traditions in Marketing*, pp. 321–342, Boston: Kluwer Academic Publishers.

KELLEY, H. H., and THIBAUT, J. W., 1978, *Interpersonal Relations. A Theory of Interdependence*, New York: Wiley.

KOGUT, B., and ZANDER, U., 1993, Knowledge of the firm and the evolutionary theory of the multinational corporation, *Journal of International Business Studies*, Fourth Quarter.

KOGUT, B., and ZANDER, U., 1995, Knowledge and the Speed of the Transfer and Imitation of Organizational Capabilities: An Empirical Test, *Organization Science*, 6, pp. 76–92.

KRACKHARDT, D., 1990, Assessing the political landscape: Structure, cognition, and power in organizations, *Administrative Science Quarterly*, 35: pp. 342–369.

MAHONEY, J. T., and PANDIAN, J. R., 1992, The Resource Based View within the Conversation of Strategic Management *Strategic Management Journal*, 13: pp. 363–80.

MAKHIJA, M. V., and GANESH, U., 1997, The relationship between control and partner learning in earning-related joint ventures, *Organization Science*, Vol. 8, No. 5 September-October.

MOORMAN, C., DESHPANDÉ, R., and ZALTMAN, G., Factors affecting trust in market research relationships *Journal of Marketing* vol. 57 (January), pp. 81–101.

MORGAN, R. M., and HUNT, S. D., 1994, The commitment-trust theory of relationship marketing *Journal of Marketing*, Vol. 58 (July) pp. 20–38.

NELSON, R. R., and WINTER, S. G., 1982, *An evolutionary theory of economic change*, Boston: Harvard Business Press.

NONAKA, I., 1994, A Dynamic Theory of Organizational Knowledge Creation *Organization Science*, 5, 1, pp. 14–37.

NONAKA, I., and TAKEUCHI, H., 1998, A theory of the firm's knowledge-creation dynamics, in Chandler, A.D., Hagström, P., and Sölvell, Ö., (eds) *The Dynamic Firm*, New York: Oxford University Press.

Oxford Advanced Learner's Dictionary of Current English, 1995, Oxford University Press.

PATEL, P., and PAVITT, K., 1998, The wide (and increasing) spread of technological competencies in the world's largest firms: a challenge to conventional wisdom, in Chandler, A.D., Hagström, P., and Sölvell, O., (eds), *The Dynamic Firm*, New York: Oxford University Press.

PERRONE, V. (ed.), 1989, *Dettagli, Orizzonti and Ingrandimenti. Osservatori* Organizzativo, CRORA, CRORA-Bocconi University, Milan.

PETERAF, M., 1993, The corner stones of competitive advantage: A resource-based view *Strategic Management Journal*, 14(3), pp. 179–191.

PFEFFER, J., and SALANCIK, G. R., 1978, *The external control of organizations. A resource dependence perspective*, New York: Harper & Row Publishers.

POLANYI, M., 1962, *Personal Knowledge: Towards a Post-Critical Philosophy*, New York: Harper & Row.

POWELL, W. W., 1987, Hybrid organizational arrangements, *California Management Review*, 30, pp. 67–87.

POWELL, W. W., 1990, Neither market nor hierarchy: Network forms of organizations, in Staw, Barry, and Cummings, L.L., (eds) *Research in Organizational Behavior*, 12: pp. 295–336. Greenwich, CT: JAI Press.

SHARMA, S., 1996, *Applied Multivariate Techniques*, Canada: John Wiley and Sons.

SNEHOTA, I., 1990, *Notes on a theory of business enterprise*, Uppsala University: Doctoral Dissertation Department of Business Studies.

SNOW, C. C., and HREBINIAK, L.G., 1980, Strategy, Distinctive Competence and Organizational Performance *Administrative Science Quarterly*, Vol. 25, June, pp. 317–336.

SZULANSKI, G., 1996, Exploring Internal Stickiness: Impediments to the Transfer of Best Practice Within the Firm. *Strategic Management Journal*, Vol. 17 (Winter Special Issue), pp. 27–43.

TEECE, D. J., PISANO, G., and SHUEN, A., 1990, Firm Capabilities, Resources and the Concept of Strategy, CCC Working Paper 90-8, CRM, U.C. Berkeley, California.

THILENIUS, P., 1997, *Subsidiary Network Context in International Firms*, Uppsala University: Doctoral Thesis no. 68, Department of Business Studies.

UZZI, B., 1996, The sources and consequences of embeddedness for the economic performance of organizations *American Sociological Review*, 61: pp. 674–698.

UZZI, B., 1997, Social Structure and Competition in Interfirm Networks: The Paradox of Embeddedness *Administrative Science Quarterly*, 42 (1997): pp. 35–67.

WINTER, S.G., 1987, Knowledge and Competence as Strategic Assets, in Teece, D., (ed.) *The Competitive Challenge: Strategies for Industrial Innovation and Renewal,* Cambridge, Mass.: Ballinger.

ZUCKER, L.G., 1991, The role of institutionalization in cultural persistence in Powell, W.W., and DiMaggio, P.J., (eds), *The new institutionalism in organizational analysis,* Chicago: The University of Chicago Press.

CHAPTER 12

Knowledge Flows in MNCs through Cross-Border and Cross-Functional Projects

KATARINA LAGERSTRÖM

Introduction

Despite the increasing globalization of most industries, research on multi-national corporations (MNCs) has concentrated more on why and how companies internationalize, and thereby become MNCs (Erramilli and Rao, 1993; Johanson and Vahlne, 1977; Reid, 1981), than on the systems and processes that MNCs use to coordinate and control their widely disparate and dispersed subsidiaries (Gupta and Govindarajan, 1994). This is so even though headquarters' main task is to control and coordinate (Cray, 1984).

Building on the work by Gupta and Govindarajan (1991, 1994) and Ghoshal and Bartlett (1990), we consider the MNC as an interorganizational network that conducts three types of inter-subsidiary transactions: (i) capital flows; (ii) product flows, such as intercorporate export flows; and (iii) knowledge flows. Knowledge flows concern the transfer of technology and/or other skills to and from various subsidiaries. Here we focus on just one of these types of transactions, namely, the knowledge flows, primarily because the proportion of more complex global MNCs is rising (Bartlett 1986; Gupta and Govindarajan, 1991). This in turn leads to knowledge flows across subsidiaries becoming particularly important in addition to capital and product flows.

An important method by which to accomplish knowledge transfer and to increase the collaboration between different units in the MNC is to bring

206

employees located in different subsidiaries together to work in creative working groups, i.e., projects. In this way they can share their specific knowledge of diverse areas. Doz *et al.* (1990), Kogut (1990) and White and Poynter (1990) have, for example, highlighted the need for large corporations to make better use of projects in different development processes to ensure that a collaborative effort is made for the key decisions and problems.

The purpose of this chapter is twofold. Its first aim is to show that cross-border and cross-functional project organizations, which are established by headquarters to achieve a specific objective that concerns larger parts of the MNC, might well result in an increase of transfer of knowledge between the subsidiaries concerned. Its second purpose is to show that the use of projects as an organizational mode may also lead to increased collaboration and, later, to the establishment of relationships between subsidiaries. A case study undertaken at the large Swedish-Swiss MNC ABB of a cross-border and cross-functional project will illustrate these suggestions.

Theoretical framework

The MNC as a network

MNCs have been classified in a variety of ways. To give three examples of ways in which one can describe and give certain characteristics of the different kinds of relationships that exist between headquarters and subsidiaries: Bartlett and Ghoshal (1989) classified MNCs according to whether they were perceived to be multinational, international, global or transnational; Hedlund's (1986; 1994) distinction was between M-formed and N-formed or heterarchical MNCs; and Porter (1986) used a spectrum that ranged from multidomestic to global. These generic global strategies are – to a large extent – consistent with Gupta and Govindarajan's (1991) discussion of different kinds of transactions undertaken between units within MNCs in three key dimensions: capital flows, product flows, and knowledge flows. The export-oriented global MNC would be characterized as being dominated by product flows from the home country to the foreign units. The multidomestic MNC would be characterized as being dominated by capital flows between the various units. Finally, according to the definition given by Gupta and Govindarajan (1991), the complex global or transnational MNC would be characterized as being dominated by a multi-directional flow of products, capital and knowledge between the subsidiaries. In this paper we have chosen to pursue a more focused in-depth discussion of issues concerning the knowledge flows between units, which is, according to Gupta and Govindarajan (1994), the most important flow in complex global MNCs.

Forsgren *et al.* (1996) also give prominence to the growing awareness that subsidiaries within the same MNC are often not alike, meaning that the relationships between the headquarters and their subsidiaries will differ. A subsidiary can be of more or less importance to the other units in the corporation, its importance being dependent on its access to resources, products, knowledge and people, and on its relationships with the actors in the local network. Subsidiaries' interdependence increases the need for the coordination of and collaboration between units. Thus managing an MNC nowadays is often said to be an attempt by headquarters to attain global integration and coordination of the dispersed subsidiaries whilst understanding the subsidiaries' need for some autonomy and to adapt to local conditions. Thus pressures to globalize and localize exist simultaneously (Bartlett and Ghoshal, 1989; Martinez and Jarillo, 1989).

Knowledge-based view and MNCs

It is widely accepted in the economic literature that, in general, knowledge can be transferred more effectively and efficiently through internal organizational mechanisms rather than through the external market. This is due to the fact that the external exchange of knowledge is susceptible to several market imperfections. As a result, MNCs will be established with the intention of internalizing the knowledge transfer (Gupta and Govindarajan, 1991, 1994; Kogut and Zander, 1992). Internalizing transactions also reduces the risks intrinsic to relying upon the market. Kogut and Zander (1993: 630) stress that it is not necessary to find the explicit motive behind or explicit reason why markets are internalized, rather, it is only of interest to conclude that firms are an efficient means by which knowledge is created and transferred across borders. The explanation they give of why firms create and transfer technology more economically is that "through repeated interactions, individuals and groups in a firm develop a common understanding by which to transfer knowledge from ideas into production and markets" (Kogut and Zander, 1993: 631). The assumption that there is a common understanding within a firm concerning the transfer of knowledge is connected to the three classical observations upon which organizational learning is built. Levitt and March (1996: 516) conclude in their literature review on organizational learning that organizational learning is viewed as a routine-based, history-dependent, and target oriented process (*cf.* Levitt and March, 1996; Huber, 1996).

An additional aspect when discussing learning and knowledge is the recognition that different types of knowledge have different implications for organizations. The most common distinction made is that between tacit and explicit knowledge (Almeida and Grant, 1998). Explicit knowledge is defined as being formal and systematic, consisting primarily of information and of factual and scientific knowledge. For this reason, it is possible to articulate knowledge that is explicit, and therefore it is easy to communicate and share (Nonaka, 1991). In contrast to explicit knowledge, tacit knowledge (or know-how) is not easily expressible since it is personal, deeply rooted in action and in the individual's commitment to a specific context, i.e., it is embodied within individuals. Tacit knowledge is also hard to formalize, and therefore difficult to communicate to others (Nonaka, 1991). We have chosen not to make a distinction between different kinds of knowledge in this paper, building on the argument made by Blackler (1995) in his article "Knowledge, Knowledge Work and Organizations: An Overview and Interpretation" that the traditional conception of knowledge – as being disembodied, individual, and formal – is unrealistic, especially if we are not interested in the character of the knowledge. Dahlqvist (1998) also argues that knowledge should not be conceptualized as an entity, but rather as a process that is intertwined with the other activities of the actors.

Gupta and Govindarajan (1991, 1994), and Kogut and Zander (1993) stress that knowledge is held by individuals, but can also be expressed in routines, through members co-operating in groups, networks and organizations. Furthermore, co-operation in different kinds of group constellations is claimed to influence and facilitate the ability to transfer knowledge within an organization because of the shared stock of technical and organizational knowledge. However, problems can arise when the knowledge is to be shifted to the rest of the organization and when it is to be put to practical use in input, throughput and output processes. Mutual adaptations between the concerned parties are a possible means of facilitating the transfer (Leonard-Barton, 1988).

One important conclusion made by Gupta and Govindarajan (1991) is that units within MNCs will differ in their engagement and therefore their contribution to the intra-corporate knowledge. The role as either provider or receiver of knowledge flows will also have an effect on the corporate control and coordination of the subsidiaries' activities. One explanation for the differences in knowledge development at a subsidiary level may be connected to relations with external actors. According to Johanson and Vahlne (1977), adaptations to environmental demands leading to differences in the knowledge base are a matter of understanding the environment.

Cross-border and cross-functional projects within MNCs

It is claimed that the competitive advantage of a whole corporation does not reside in a few specific units, but in MNC's ability to identify, exploit and spread knowledge amongst its subsidiaries. This makes it essential for MNCs to develop a structure that is flexible, i.e., adopt new organizational modes and adopt a new organic structure (Bartlett, 1986; Bartlett and

Ghoshal, 1989; Lorange and Probst, 1990). Mintzberg (1983) stresses that these structures need to break through the conventional boundaries of specialization and differentiation, to gather together those with the requisite knowledge, by, for example, gathering specialists in multidisciplinary project teams, each formed for a specific assignment. Bartlett and Ghoshal (1989) also argue that the new organizational forms have to make greater use of mechanisms that might be called microstructural tools, one such tool being the initiation of projects. A general definition is that a project can be perceived to be the gathering of human and non-human resources in a temporary organization that cuts across ordinary organizational borders and that can vary greatly in terms of its size and assignment (Cleland and King, 1968; Wilson, 1994).

Like other projects that do not fall within the framework of a permanent organization, the organizations put in place for cross-border and cross-functional projects are usually characterized by three features. The first is that a project has a *single specific assignment/objective* that is to be completed, i.e., it has a formalized purpose that is of importance to the entire corporation (Miles, 1965; Anthony *et al.*, 1965). The second feature is that a project has a *specific start and end date* (Kerzner, 1984), which is the most obvious difference between projects and permanent organizations. While a permanent organization is presumed to endure, a project has a formal end date after which it ceases to exist (Gaddis, 1959; Packendorff, 1993). It is, however, possible to see this feature from two different angles. The first one, often represented in the project management literature, is the end date specified at the start of the project. This end date is then essentially a matter of delivery and the project will come to an end whether or not the objective is reached. According to the second angle, the project team will be disbanded on completion of the assignment, i.e., there is no definite end date for the project, instead the project finishes once the objective is reached. The third feature for a project is that every project has a *specified budget*, that is, the resources needed for the assignment are decided at the outset. This feature is not usually expressed explicitly, but tends to be an underlying criterion mentioned when the objective of a project is laid down, through such phrases as to "reach the objective of the project within the specified time and budget."

Organizing cross-border and cross-functional projects can be afflicted with difficulties since the subsidiaries may look upon the project as something that is imposed on them by headquarters. This could particularly be the case for subsidiaries that have become highly autonomous and have stronger connections to their local network than with other units within their own corporation (*cf.* Forsgren *et al.*, 1996; Forsgren and Pahlberg, 1992; Andersson, 1997). In addition, the subsidiaries must be prepared to appoint employees to participate in the team working on the project and sometimes even to establish a special position for an employee. This takes both human and financial resources.

Another aspect of cross-border and cross-functional projects initiated by headquarters is that they may fulfill other important roles in the MNC than the official assignment. Martinez and Jarillo (1989; 1991), for example, stress that among other administrative mechanisms, lateral relations formed through task forces or projects, that cut across the formal structure, are a subtle mechanism with which headquarters in particular can achieve integration of the different units' activities in the MNC. Doz and Prahalad (1992) state that task forces and intraorganizational projects are mechanisms used by headquarters to gain control over the subsidiaries' operations; they are also considered to be instruments for short-term use used to obtain quick results. Projects may also play a third role within MNCs according to, for example, Bartlett and Ghoshal (1989), Gupta and Govindarajan (1994) and Mintzberg (1983), namely, to facilitate and increase the transfer of knowledge in terms of skills and expertise between the various subsidiaries. Mintzberg (1983) even claims that organizing work in teams comprised of people with different experience to offer from different areas is vital for developing new knowledge and skills within large corporations.

From the theoretical discussion, one can expect to find that cross-border and cross-functional projects established by headquarters are of importance for multinational corporations when there is a particular problem to solve or when a solution is to be implemented that has implications for subsidiaries in more than one country. It is also possible to argue that a well-defined project with a well-expressed assignment will be more efficient when it comes to gathering the relevant data and obtaining the information necessary for the completion and implementation of the assignment than either involving the whole of the permanent organization would be (*cf.* Prescott and Smith, 1987), or allowing a single unit in the corporation – regardless of its hierarchical level – to be responsible for undertaking the assignment. Studies have shown that there is a risk of the so-called Not-Invented-Here Syndrome (NIH) (Bartlett and Ghoshal, 1989: 119) appearing, meaning that the internal acceptance of a solution could be very low if those subsidiaries that will be affected by the project are not included in the process.

The organizational structure of the whole corporation is, as argued above, also of significance when it is both feasible and appropriate to create projects for certain assignments. But the organizational structure of the project is at least equally important if the project is going to be able to take advantage of being a cross-border and cross-functional operation that can utilize the knowledge and skills located at the different subsidiaries in countries anywhere in the world. A poor organization and poor project planning are not only hazardous for the project itself, but also for the corporation as a whole since many valuable resources, both financial and human, will be tied up for the duration of the project.

On the basis of the theoretical discussions of MNCs' structures and of knowledge flows, it would be possible to pursue the idea that the efficient completion of the specific assignment is not the only objective when projects are created, even if it is the most obvious one. Instead, projects can be used as a means of facilitating and increasing the transfer of information and skills between subsidiaries. However, there are likely to be a number of factors that will hinder an efficient transfer of knowledge. People, for example, do not always want to make their knowledge accessible to other members of a project since there is a risk that they may lose the opportunity to exploit the knowledge themselves, or because others could look better than them if they get access to the knowledge (*cf.* Kostova, 1999). There is also a likelihood that members of the project do not know whom to transfer information to, or who possesses the kind of information they need. One way to facilitate the transfer could be if the cross-border and cross-functional project has a distinct and substantiated structure that is well known to all concerned.

It is also reasonable to assume that organizing work in the form of projects is a good method for establishing relationships between individuals that will last longer than the project itself since cross-border and cross-functional projects are built on relations between employees from subsidiaries in different countries around the world. Building on the work by Mintzberg, it is also possible to argue that the mutual adjustments that have to take place within a project that involves representatives with different backgrounds who have to reach a common solution that meets their different demands could be one of the prime coordination mechanisms in multinational corporations with dispersed subsidiaries (Mintzberg, 1983).

Research method

The method used here for studying cross-border and cross-functional corporate projects within large MNCs is a mixture of qualitative and quantitative methods, so-called triangulation (Jick, 1979). The questions raised are first investigated through an intensive case study of the environmental management project at the Swedish-Swiss corporation ABB, which is to be discussed in this paper. This case can probably be considered to be an atypical one (Hägg and Hedlund, 1979) since the objective of the project under study differed in terms of the kind of work normally undertaken within the frameworks of project organizations in large MNCs and in terms of the forms of the projects previously studied in business research.

The object of the study, the environmental management project, meets the three main characteristics for cross-border and cross-functional projects that take place within the framework of a permanent organization. The objective of the ABB project was to make sure that a large number of the production and service units within the corporation worldwide had at least started to implement the ISO 14001 system before the official end date set for the project, i.e., before the end of 1998. The project was coordinated by the central unit ABB Corporate Staff – Environmental Affairs, which of course had a plan and a budget for the work to be performed. With most large projects, the results or the solutions of the project have to be integrated with the ordinary business operations during or after the work has been performed in the project. Such was the case for the environmental management project at ABB. This means that in some ways the project will live on, even though it will have come to an end officially.

In the ABB case study, interviews were conducted at the central unit, ABB Corporate Staff – Environmental Affairs, which is a special unit created by the CEO and corporate management to be in charge of implementing the environmental management system at the production and service sites within the corporation. Interviews were also conducted with the people responsible for the individual countries' implementation of the environmental management system (the "Country Environmental Controller") and with employees responsible for their own unit's system (the "Local Environmental Controller") at the selected subsidiaries. The interviewees selected were people with considerable knowledge of the environmental project, but also with special in-company knowledge. All the information conveyed was carefully compared with other available data coming from annual reports, company records, etc., with a view to increasing the credibility of the study (cf. Hägg, 1981).

The interviews were semi-structured. That is, the questions the respondents were asked were mainly those that it was considered would give a reasonably good background to and understanding of how large corporate projects are organized and managed for this special case. Some questions, however, were not only intended to gain evidence on the concrete aspects of the project work, but instead, they also sought to reveal personal subjective information from each individual interviewed. This was to gain information about his or her experiences of working on a large cross-border and cross-functional project, i.e., of being involved in a project built not only on strong lateral relations between production and service subsidiaries in one country but also between sites in different countries.

Empirical data

Since the 1980s the large Swedish-Swiss MNC ABB, with sites in more than 140 countries, has been working with the task of becoming both global and local. That is, of being able to meet local demands from customers and other actors, whilst taking advantage of global core technologies and global economies of scale. Researchers have considered ABB to be one of very few corporations that have succeeded with this agenda, and perhaps the corporation can be said today to correspond to what in theoretical terminology is called a complex global or transnational organization. To handle these conflicting demands, the corporation has used different methods, such as the transfer of managers between local sites, the setting up of committees and the creation of projects.

The case below is an illustration of how ABB, which has a rather complex matrix structure, has chosen to organize and manage a corporate cross-border and cross-functional project with more than 40 countries involved. That is, how it is implementing its environmental management system. The project was initiated by corporate management and is considered to be a project of strategic importance for the whole corporation. Thus it is significant that all the sites concerned around the world should have an approved system implemented before January 1999, the official end date decided for the project by the corporate management and the CEO.

The necessity of a cross-border and cross-functional environmental management project at ABB

The environmental activities in business corporations are often considered to involve strategic decisions. Environmental awareness is increasing among firms and a couple of international standards for environmental management systems have been developed. These systems are now beginning to be implemented in industries worldwide by using one of a number of different procedures. The most well known systems are EMAS (Eco Management and Audit Scheme) and ISO 14000, the standards of the International Organization for Standardization. The systems provide the framework for the environmental work in corporations, but within these frameworks, corporations are reasonably free to implement a system adapted to their own business operations. There are also country specific circumstances that influence the outline of the system at different sites within the same corporation. Naturally the most suitable implementation is required, so these country specific circumstances will influence the way that the system is executed in large MNCs. Apart from the differences between countries that corporate management must consider when planning the organization and management of a corporate assignment, it must also take the organizational structure of the corporation into account.

At the beginning of the 1990s it was decided that ABB should develop and implement an environmental management system at 583 production and service sites in the MNC. There were already a couple of employees within the corporation who had been involved in environmental issues, both within and outside the corporation, but without being given any corporate responsibility. A special unit – Corporate Staff – Environmental Affairs (CS-EA) – was now made responsible for coordinating and organizing the implementation of the environmental management system within the MNC. This was done even though it was not in accordance with a former policy, i.e., with the decentralization process that had made subsidiaries responsible for their activities. A primary reason for the corporate management deciding to make a special unit responsible for the process was the considerable difference in environmental awareness in the subsidiaries, which could be explained by the organizational structure of ABB.

ABB is an organization that is highly decentralized, and in which all subsidiary managers are fully responsible for their own subsidiaries and for all the activities performed within them. The decentralization process during the 1980s and 1990s that created these autonomous subsidiaries resulted in the loss of knowledge in different areas, of which the environmental area was one. Before the decentralization, environmental issues had been a prioritized corporate matter, with high demands being imposed on all the subsidiaries. Besides that, a lot of competence was lost in this area because of the decentralization, and it became the responsibility of each subsidiary manager to handle environmental issues in whatever way he or she thought most suitable. This resulted in some subsidiaries not participating in this fast developing area and, in principle, completely lacking any knowledge of environmental issues when the corporate management decided that the sites should have an environmental management system implemented before the beginning of 1999. Other subsidiaries who had continued the previously introduced corporate agenda – as a result of requests from customers, or because of a general interest in environmental issues or high legal demands in their home country – were now in possession of knowledge that was of importance to the whole corporation. Therefore one of the corporate unit's most important tasks is the creation of an organization that can make sure that the environmental knowledge some subsidiaries possess is made available in an appropriate form to subsidiaries who lack the necessary knowledge to implement an environmental management system.

Organizing a cross-border and cross-functional project at ABB

The corporate project is organized in three layers, that is, the project consists of Corporate Staff – Environmental Affairs (CS-EA) and networks of members at both the country and the subsidiary level (see Figure 12.1). The environmental project dovetails closely with ABB's main organizational structure since it is a project where both Country Environmental Controllers (CECs) and Local Environmental Controllers (LECOs) hold

FIGURE 12.1
The organization of the Environmental Management Project at ABB

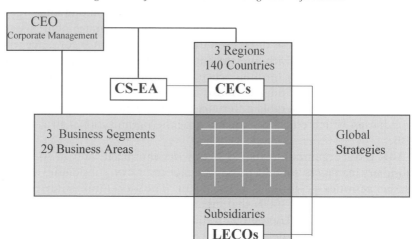

dual positions in the organization. The holding of dual positions creates some difficulties, especially in countries where the duties of CECs and LECOs are considered to be less important than their other obligations. This is mainly because the environmental awareness is low in these countries, and because there is hardly any pressure today from external actors, such as customers and the government. An example of the difference attributed to these two roles is that the work conducted for the environmental project is not included as one of the key performance variables in the bonus system according to which the CECs' and LECOs' performance is evaluated.

The special corporate unit CS-EA is established by headquarters to be responsible for developing, communicating and implementing the environmental management program at the production and service sites. The CS-EA unit reports directly to the corporate management and the CEO and produces the annual environmental management report. The two people in the CS-EA unit divide the work into two main spheres with each of them taking responsibility for one sphere. The first is to function as a service or support unit for the employees put in charge of the implementation of the environmental management system at the country and subsidiary levels. This also implies the responsibility to arrange the necessary education for CECs, and occasionally also for LECOs, on the environmental management system. The education is free of charge since all education and consultation the CECs and LECOs require is provided at the expense of the CS-EA unit, i.e., headquarters. The second sphere is to act as a lobbying unit, trying to interest the country and business area managers in, and make them aware of, the necessity to begin to introduce the system at their sites. In spite of the fact that the CS-EA unit can be regarded as the project leader for the whole environmental management program, it is not officially in charge of appointing CECs and LECOs.

The country management appoints the CECs and there is no demand from the corporate unit CS-EA that the assigned employee has prior experience or prior knowledge of the environmental management system. It is more important that the CEC has had an extensive career within the corporation and has a genuine interest in environmental issues. In some countries that are very far behind environmentally, it has turned out to be a difficult task to find employees with an interest in or even an awareness of the importance of environmental issues. This is especially problematic since the CECs in these countries often need to play an even more important role than the other CECs. This is partly because they need to persuade the management at the local subsidiaries to get started – and the management tends to think that there are other more urgent issues – and partly because they have to act as the provider of knowledge to LECOs at the subsidiaries.

As well as being responsible for the implementation of the system, the CEC must also see to it that the LECOs are given the necessary education and all help required to run the project at their local subsidiary. The CEC is also responsible for arranging the yearly meetings for all the LECOs and for organizing and sending the CS-EA unit the obligatory documentation from all subsidiaries concerning the environmental development at each site. Apart from these monthly reports and the annual official presentation in "Country Environmental Controllers' Status Report" for the environmental work at ABB, the CEC also has ongoing contact with the country management.

The subsidiary managers usually assign an employee, often with some prior experience of both project work and quality control, or with a special interest in environmental issues, to the LECO post. In some cases the CS-EA and CEC may use their influence to affect the choice of LECO, but like all other matters concerning the subsidiary, issues connected to the environmental project are officially a local management decision. The choice of LECO does not seem to be too difficult at the subsidiaries in most countries, but in a few cases there are some difficulty finding suitable employees with the relevant background. One reason for this is that knowledge of environmental issues is scanty amongst the employees at the sites and even more in certain countries on a general level. There is also a problem associated with the very same subsidiaries or countries perceiving that other matters, such as safety or quality control, are more pressing, and that these must be dealt with before it is even possible to begin to consider the implementation of an environmental management system. That the sites does not have implemented a quality control system, such as ISO 9000, yet is obviously problematic since most respondents mean that the first step in the process to implement an environmental management system, which is considered as the second or third step, is to have a working quality control system. Experiences made during the implementation of ISO 9000 are considered by respondents to be of significance when implementing ISO 14001 partly because the systems resemble each other, partly because the employees will then have a prior understanding of implementing and working with a similar system.

LECOs are responsible for implementing the system in their subsidiaries and during the process they are usually in continuous contact with CEC to get all the necessary education and help. At some subsidiaries, particularly if that site is the very first in that country, the CS-EA unit occasionally participates actively in the process. The most common way to organize the implementation of the environmental management system is to create local project groups at each site with strong connections between them and with support from the CEC.

The organization and the primary duties of the three different layers in the corporate project has been described well by a CEC, who said: "CS-EA's main task is to transmit knowledge about the environmental management system to the CECs who, in their turn, should coordinate the environmental work on a country level and transmit knowledge and the philosophy of the environmental agenda to the LECOs."

Sharing knowledge and information

The implementation of the environmental management system at ABB started with 15 pilot projects, i.e., at 15 subsidiaries around the world. This group consisted mainly of those subsidiaries that had continued working with environmental issues over the years. These first sites received a considerable amount of support from their respective CECs, but perhaps the most significant actor was the CS-EA unit, which played a very important role in several ways.

The employee from the CS-EA unit who participated actively in the implementation process at the subsidiaries was one of the company representatives in the international delegation that assisted with the development of the ISO 14000 standard on which ABB's environmental management system is built. It is also according to this standard that the subsidiaries' systems get certified by special external firms, who, by the way, also participated in the delegation. Since the ISO 14001 system was under development when ABB began to introduce the environmental management system, the ABB representative could influence the shaping of the system to some extent and was also in a position to transfer first hand information to the LECOs about the standard and the continuous changes being made to it.

The uncertainty in the ISO 14001 system was rather high in the beginning, and not only did the companies that were to implement it feel the need for support, but the certifying firms too perceived that they did not have enough experience of ISO 14001 in general. In some countries this led to a few of these certifying firms taking an active part in the whole implementation process at a few of the local ABB sites. One could say that the certifying firms and representatives from ABB's environmental management project at different levels shared their antecedent knowledge and worked together to create a suitable method for and reasonable demands on companies who were going to implement an environmental management system in the future.

The CS-EA unit was a very significant actor in the beginning, because it collected, arranged and made the knowledge the subsidiaries possessed before the process started and what is continuously learnt during the process available to other CECs and LECOs in the corporation. This is done by means of manuals written by the CS-EA unit, and through making each of the subsidiaries' own documentation accessible to the other sites. The corporate unit also require that the subsidiaries with a working system host an internal education course for the newly appointed CECs, and also occasionally the LECOs, in the corporation. That the education normally takes place at a certified site gives the CECs the opportunity to speak to the LECO and other employees who have experience of the whole process. These courses were also mentioned by one CEC as providing a good opportunity for the formation of supportive relations between the participating CECs.

Besides the education the CS-EA unit organizes, it also arranges and attends the annual workshop for CECs. These workshops or conferences are partly a tool for creating a foundation from which the CECs can form relationships amongst themselves, and are partly an easy means by which the corporate unit can simultaneously inform almost all CECs about changes in the demands in the ISO 14001 system or in other areas relevant to the environmental work.

The workshop also provides the CECs with an opportunity to share the experiences they have had when implementing the system and what they have learnt from one another's achievements and mistakes. According to the CECs, these meetings are helpful when there is a need for them to get in contact with CECs who have experience in certain environmental sectors, usually as a result of their having encountered a problem. One CEC mentions that one of the LECOs ran into a rather complex asbestos problem that had to be solved before the site could get its system approved. Since neither the LECOs in that country, nor the CEC, had any prior experience in this area, they needed to get help from elsewhere. At one of the annual workshops the CEC had crossed paths with another CEC who had mentioned that one of his/her LECOs had had a similar problem. This relation was now used for getting the necessary help to solve the problem. The two CECs acted in this special case as intermediaries, but in other cases where the LECOs are from the same country, or when there are no language difficulties, a CEC may just make sure that the LECOs get in contact with each other and then leave them to solve the problem by themselves. The obligatory annual meeting arranged by the CECs for all the LECOs in one country, appears to be a good way of facilitating direct contact among the LECOs in a country. The purpose of the workshops as the LECOs see it is very much the same as the purpose of the CEC workshops: namely, to make the two networks function on their own. That is, to make sure that the actors get the opportunity to establish relationships and use them to solve problems without receiving help from one level higher up the hierarchy, i.e., from the CS-EA unit or the CEC.

The corporate unit is also responsible for a computer-based communication system, an intra-net system, through which the CECs and LECOs can communicate and share their experiences of working with ISO 14000. Besides facilitating the sharing of experiences, the communication system fulfills the objective of informing the CS-EA unit of the progress in the different countries. Some CECs only use the intra-net to inform the CS-EA unit of developments, but they are nevertheless of the opinion that it is probably a useful tool for increasing information transfer, because it is most likely that from time to time different CECs run into similar difficulties. If the CECs present their solutions on the intra-net, they are made accessible to other actors in the project. There is one problem associated with the use of the intra-net system as one of the main sources for sharing information, namely, that in some countries the LECOs do not have any knowledge of the system, and in some cases it would not help even if they knew about it since they do not have a computer at their disposal or know how to use the computer or the intra-net.

Sharing experiences and solutions to problems is probably the most important way to ease the implementation process in a corporation where the different subsidiaries possess knowledge related to different areas.

External actors in a cross-border and cross-functional project at ABB

Opinions among the actors in ABB's environmental management project differ to some extent when it comes to the matter of whether to involve external actors, such as consultants, in the implementation process. The CS-EA unit is of the opinion that there is sufficient competence and experience within the corporation, making it unnecessary for either CECs or LECOs to engage consultants. This does not, however, imply that the sites concerned are not allowed to select whatever solution suits them the best.

The result is a non-uniform solution in the countries concerned and even from one subsidiary to another in the same country.

Today consultants are used rather extensively at both the country and the subsidiary level. It also appears that the CECs generally take an active part in assisting the subsidiaries to select a consulting firm, and moreover, are involved in decisions, such as how much time and during which phases consultants should be engaged in the implementation process at the subsidiaries. One of the CECs stresses that it was almost inevitable that the subsidiaries would involve consultants because of the fast implementation and the LECOs' restrictions on their time. The CEC also emphasized that he would not have enough time to help all the subsidiaries individually, and instead he has decided to focus on coordinating the work and to occasionally assist certain subsidiaries.

Another group of external actors of importance to all subsidiaries involved in the project is the group of firms that perform the certification of the ISO 14001 system. There are a few certification firms to choose from, but it seems as though the decision to hire one firm is built on prior experience and on existing relationships between a particular firm and ABB as a corporation or with specific subsidiaries in each country. Quite a few of the subsidiaries chose to use the same firms as were hired before to perform the certification of their quality system ISO 9000.

Concluding remarks

Comments on a cross-border and cross-functional project at ABB

ABB headquarters' decision to create a cross-border and cross-functional project for the implementation of the environmental management system at the 583 production and service subsidiaries worldwide appears to have been a wise decision for several reasons. The most important reason was probably that by creating a project, the different subsidiaries' knowledge about environmental issues became accessible to other actors within the corporation. This was especially important in this case as the knowledge possessed by the subsidiaries varied to a large extent, primarily because of differences in external demands from customers and other actors in the subsidiaries' local networks. This led to those subsidiaries who had continued the environmental agenda within ABB now becoming the providers of knowledge, and not only to other subsidiaries in the corporation that were lagging on environmental issues, but also to external actors involved in the project.

The importance put on matters concerning the transfer of knowledge is also reflected in the organizational structure of the project, namely, the three-layer structure comprised of the special unit, Corporate Staff – Environmental Affairs, and the two networks of members at a country and subsidiary level. The most important task for the CS-EA unit is to facilitate and enhance the transfer of knowledge between the sites. Since this transfer of knowledge is difficult, even impossible according to some researchers, the unit has focused on two kinds of methods to make sure that at least general information is transferred among the participating countries and their local sites. The first consists of various kinds of printed matter, for instance, the CS-EA unit has compiled a manual that gives the CECs and LECOs instructions and advises them how to implement the system. In addition, the subsidiaries, and especially those 15 pilot subsidiaries, are obliged to make available their documentation of the work performed on behalf of the project. The way in which these subsidiaries chose to organize and arrange their implementation has actually been taken as a role model by the other sites in the corporation. Another means developed by the CS-EA unit to facilitate the transfer of knowledge is the creation of an intra-net system. The second method used by the corporate unit is different arrangements, which enable the CECs and LECOs to meet in person and gives the opportunity to establish relations. The two main arrangements of this type are the education of newly appointed CECs on the environmental management system at local ABB sites where systems have already have been implemented, and the annual workshops for the CEC and LECOs. Of course the special unit CS-EA also plays a significant role since one of the members visits and participates in the implementation at certain subsidiaries. Another reason for headquarters to establish a cross-border and cross-functional project, which is partly connected to the reason discussed above, is that the implementation of the system is considered a strategically important matter, which therefore has to incorporate sites in each country in which ABB is represented.

Comments and further research

The case study of the environmental management project at ABB gives good reason to assume that cross-border and cross-functional projects are not only established by headquarters to solve a particular assignment in an efficient way, but that they also provide a means by which to increase and facilitate the transfer of knowledge between different subsidiaries in the MNC. This is particularly the case if there is a large difference in the sites' knowledge. An additional motivation for using this kind of organizational mode is that the relationships established within the framework of the project seem to result in collaborations between the participating subsidiaries, beyond the direct objective of the project. This is an aspect that is stressed by, for example, Martinez and Jarillo (1989; 1991) in their theoretical discussion about the utilization of different administrative mechanisms.

How well headquarters succeed in creating a cross-border and cross-functional project that leads to transfer of knowledge and to establishment of relations between the subsidiaries is probably contingent upon the degree of dependence of resources, products, people and knowledge between the participating actors in the project. An additional aspect that probably has a considerable influence on how successful the project is in meeting the corporate objective is the structure of the project, i.e., how the work is organized and managed. This does not, however, imply that there is only one way to organize a cross-border and cross-functional project. Rather, it is probably advisable to create a structure for that specific project that makes it conceivable for the actors to meet and communicate directly with one another. Another aspect connected to the structure of the project and the creation of relationships between actors in the project that has not been discussed explicitly in this paper, but must be of interest, is trust. This needs to be studied further. According to Morgan and Hunt (1994), co-operation to achieve mutual goals promotes the establishment of relationships, which in turn influences the commitment of the actors involved and the trust between them. In their research, Moorman, Zaltman and Deshpandé (1992) have been able to show that information provided by a trusted party is used more often than that from an unknown entity, and thereby provides greater value to the recipient. Presumably this reasoning is also applicable to the circumstances concerning relationships and the transfer of knowledge between the participating subsidiaries in large cross-border and cross-functional projects in multinational corporations.

References

ALMEIDA, P., and GRANT, R. M., 1998, International Corporations and Cross-Border Knowledge Transfer in the Semiconductor Industry, A Report to the Carnegie-Bosch Institute.

ANDERSSON, U., 1997, *Subsidiary Network Embeddedness: Integration, Control and Influence in the Multinational Corporation*, Uppsala University: Department of Business Studies, (Ph.D. diss.).

ANTHONY, R. N., DEARDEN, J., and BEDFORD, N. M., 1965, *Management Control Systems*, Richard D. INC., USA.

BARTLETT, C. A., 1986, Building and Managing the Transnational: The New Organizational Challenge, in Porter, M. (ed.), *Competition in Global Industries*, Boston: Harvard Business School Press.

BARTLETT, C. A., and GHOSHAL, S., 1989, *Managing Across Borders: The Transnational Solution*, Chatham, Kent: Mackays of Chatham PLC.

BLACKLER, F., 1995, Knowledge, Knowledge Work, and Organizations: An Overview and Interpretation *Organization Studies*, Vol. 16, pp. 1021–1046.

CLELAND, D. I., and KING, W. R., 1983, *System Analysis and Project Management*, New York: McGraw-Hill.

CRAY, D., 1984, Control and Coordination in Multinational Corporations *Journal of International Business Studies*, Fall, pp. 85–98.

Katarina Lagerström

DAHLQVIST, J., 1998, *Knowledge use in Business Exchange: Acting and Thinking Business Actors*, Uppsala University, Department of Business Studies, (Ph.D. diss.).

DOZ, Y., and PRAHALAD, C. K., 1992, Headquarters' Influence and Strategic Control in MNCs, in Bartlett, C. A., and Ghoshal, S., (eds), *Transnational Management: Text, Cases and Readings in Cross-border Management*, Irwin.

DOZ, Y., PRAHALAD, C. K., and HAMEL, G., 1990, Control, Change, and Flexibility: the Dilemma of Transnational Collaboration, in Bartlett, C. A., Doz, Y., and Hedlund, G., (eds), *Managing the Global Firm*, New York: Routledge.

ERRAMILLI, M. K., and RAO, C. P., 1993, Service Firms' International Entry-Mode Choice: A Modified Transaction-Cost Analysis Approach, *Journal of Marketing*, Vol. 57, July, pp. 19–38.

FORSGREN, M., and PAHLBERG, C., 1992, Subsidiary Influence and Autonomy in International Firms, *Scandinavian International Business Review*, Vol. 1, No. 3, pp. 41–51.

FORSGREN, M., PAHLBERG, C., and THILENIUS, P., 1996, Culturally Induced Problems and Control in MNCs, in Pahlberg, C., *Subsidiary – Headquarters Relationships in International Business Networks*, Uppsala University, Department of Business Studies, (PhD. diss.).

GADDIS, P. O., 1959, The Project Manager *Harvard Business Review*, Vol. 37, May-June, pp. 89–97.

GHOSHAL, S., and BARTLETT, C. A., 1990, The Multinational Corporation as an Interorganizational Network *Academy of Management Review*, Vol. 15, No. 4, pp. 603–625.

GUPTA, A., and GOVINDARAJAN, V., 1991, Knowledge Flows and the Structure of Control within Multinational Corporations *Academy of Management Review*, Vol. 16, No. 4, pp. 768–792.

GUPTA, A., and GOVINDARAJAN, V., 1994, Organizing for Knowledge Flows within MNCs *International Business Review*, Vol. 3, No. 4, pp. 443–457.

HÄGG, I., 1981, Validering och generalisering – problem i företagsekonomisk forskning (Validation and Generalization – a Problem in Business Research), in Brunsson, N. (ed.) *Företagsekonomi – sanning eller moral?* Studentlitteratur, Lund.

HÄGG, I, and HEDLUND, G., 1979, Case Studies in Accounting Research *Accounting, Organization and Society*, Vol. 4, No. 1/2, pp. 135–143.

HEDLUND, G., 1986, The Hypermodern MNC – A Heterarchy? *Human Resource Management*, Vol. 25, No. 1, pp. 9–35.

HEDLUND, G., 1994, A Model of Knowledge Management and the N-form Corporation *Strategic Management Journal*, Vol. 15, pp. 73–90.

HUBER, G. P., 1996, Organizational Learning: The Contributing Processes and the Literatures, in Cohen, M. D., and Sproull, L. S., (eds) *Organizational Learning*, California: Sage Publications, Inc.

JICK, T. D., 1979, Mixing Qualitative and Quantitative Methods: Triangulation in Action in Van Maanen, J., (ed.), *Qualitative Methodology*, London: Sage Publications.

JOHANSON, J., and VAHLNE, J-E., 1977, The Internationalization Process of the Firm – A Model of Knowledge Development and Increasing Foreign Market Commitment *Journal of International Business Studies*, Vol. 8, Spring/Summer, pp. 23–32.

KERZNER, H., 1984, *Project Management: A Systems Approach to Planning, Scheduling and Controlling*, New York: Van Nostrand Reinhold Co.

KOGUT, B., 1990, International sequential advantages and network flexibility, in Bartlett, C. A., Doz, Y., and Hedlund, G., (eds), *Managing the Global Firm*, New York: Routledge.

KOGUT, B., and ZANDER, U., 1992, Knowledge of the Firm, Combinative Capabilities, and the Replication of Technology, *Organization Science*, Vol. 3, No. 3, August, pp. 383–397.

KOGUT, B., and ZANDER, U., 1993, Knowledge of the Firm and the Evolutionary Theory of the Multinational Corporation *Journal of International Business Studies*, Fourth Quarter, pp. 625–643.

KOSTOVA, T., 1999, Transnational Transfer of Strategic Organizational Practice: A Contextual Perspective, *Academy of Management Review*, Vol. 24, No. 2, pp. 308–327.

LEONARD-BARTON, D., 1988, Implementations as Mutual Adaptations of Technology and Organization *Research Policy*, Vol. 17, pp. 251–267.

LEVITT, B., and MARCH J. G., 1996, Organizational Learning, in Cohen, M. D., and Sproull, L. S., (eds), *Organizational Learning*, California: Sage Publications, Inc.

LORANGE, P., and PROBST, G., 1990, Effective Strategic Planning Processes in the Multinational Corporation, in Bartlett, C. A., Doz, Y., and Hedlund, G., (eds), *Managing the Global Firm*, New York: Routledge.

MARTINEZ, J. I., and JARILLO, J. C., 1989, The Evolution of Research on Coordination Mechanisms in Multinational Corporations *Journal of International Business Studies*, Fall, pp. 489–514.

MARTINEZ, J. I., and JARILLO, J. C., 1991, Coordination Demands of International Strategies *Journal of International Business Studies*, Vol. 22, No. 3 , pp. 429–444.

MILES, M. B., (ed.), 1965, *On Temporary Systems*, New York: Teachers College Press.

MINTZBERG, H., 1983, *Structure in Fives: Designing Effective Organizations*, New Jersey: Prentice-Hall Inc.

MOORMAN, C., ZALTMAN, G., and DESHPANDÉ, R, 1992, Relationship Between Providers and Users of Market Research: The Dynamics of Trust Within and Between Organizations *Journal of Marketing*, Vol. 29, August, pp. 314–328.

MORGAN, R. M., and HUNT, S. D., 1994, The Commitment-Trust Theory of Relationship Marketing *Journal of Marketing*, Vol. 58, July, pp. 20–38.

NONAKA, I., 1991, The Knowledge-Creating Company *Harvard Business Review*, Nov-Dec, pp. 96–104.

PACKENDORFF, J., 1993, *Projektorganisation och Projektorganisering – Projektet som plan och temporär organisation (Project Organization and Project Organizing – The Project as a Plan and as a Temporary Organization)*, FE-publikationer 1993: No. 145, Handelshögskolan in Umeå (Ph.D. diss.).

PORTER, M., 1986, Competition in Global Industries: A Conceptual Framework, in Porter, M., (ed.), *Competition in Global Industries*, Boston, Mass: Harvard Business School Press.

PRESCOTT, J. E., and SMITH, D. C., 1987, A Project-Based Approach to Competitive Analysis *Strategic Management Journal*, Vol. 8, pp. 411–423.

REID, S., 1981, The Decision-Maker and Export Entry and Expansion *Journal of International Business Studies*, Fall, pp. 101–112.

WHITE, R. E., and POYNTER, T. A., 1990, Organizing for world-wide advantage, in Bartlett, C. A., Doz, Y., and Hedlund, G., (eds), *Managing the Global Firm*, New York. Routledge.

WILSON, T. L., 1994, Off-Project Management: Implications from Observations of Product Development Projects, in Lundin, R. A., and Packendorff, J., (eds), *Proceedings on the International Research Network on Organizing by Projects (IRNOP) Conference on Temporary Organizations and Project Management. 22–25 March, 1993*, in Lycksele, Handelshögskolan in Umeå, Umeå.

CHAPTER 13

Management Control Systems: A Tool for Learning in the Global Company

CARIN B. ERIKSSON and JAN LINDVALL

Introduction

Two important trends are emerging within highly internationalized firms. The first involves change programs, in which the traditional organizational structure is being questioned at the same time as a new structure is being developed. This trend has often been described as a change from a hierarchical structure to a network-oriented structure (Buckley and Casson, 1998, p. 31). The second trend involves criticism of traditional management control systems, and the simultaneous development of a new, modern method of management control (Bromwich and Bhimani, 1994). Both trends aim to improve the conditions for organizational learning, for it is accepted that management control systems are an important tool for organizational learning (Friso den Hertog, 1978). The purpose of this article is to discuss whether business firms with a network-oriented structure and with greater needs for coordination make use of a management control system that increases opportunities for organizational learning.

Theoretical framework

New structure

The need to adapt to local circumstances has meant that the international activities of many business firms have been characterized by a great deal of independence on the part of different national companies. This is changing, and the aim now is to achieve coordination and a more comprehensive identity. Independence created the risk of local "kingdoms" within a business firm: now it is essential to appear as *one* company to global customers. This greater emphasis on coordination has led to the appearance of an organizational form with many names, including heterarchy (Hedlund, 1993), integrated network (Bartlett and Ghoshal, 1989), or holarchy (Sjöstrand, 1997). One distinctive feature of the new organizational form is that the company is no longer to be perceived as having *one* center. Gone is the perception that the firm's international subsidiaries are controlled from the centrally situated parent company's headquarters (Forsgren, 1988; Pasternack and Viscio *et al.*, 1998). The new structure is characterized by many strong units. These strong, local units call in question the traditional equal treatment of subsidiaries (Bartlett and Ghoshal, 1989, p. 99). Instead, units may be given a special global responsibility within the group in regard to production, research and development, or, more common lately, administration (CFO Europe, 1998). These units are often called centers of excellence, international product centers, or shared service centers.

Another distinctive feature of the new organizational form is that the business is carried on in many different geographical areas (Dunning, 1998). This makes it possible to take advantage of specific countries' competitive advantages, such as lower factor costs or greater possibilities for learning and innovation. The growth of dispersed and specialized assets also explains the increase in interdependent relationships. Bartlett and Ghoshal (1989, p. 92) express this as follows:

> Today's worldwide competitive environment demands collaborative information sharing and problem solving, co-operative resource sharing and collective implementation – in short, a relationship built on interdependence.

This interdependence is the most interesting and distinctive feature of the new organizational form.

New management control systems

The need for more coordination within organizations concerns everything from the relatively simple coordination of the physical flow of goods to the more complicated distribution of information (Bartlett and Ghoshal, 1989, p. 170). Management control systems play an important role in the latter. As Anthony and Govindarajan (1995, p. 10) express it:

> The system (management control system) is built around a financial core .... The reason for a financial core is that the system is partly a coordinating device, and money is the only common denominator that can

be used to measure, summarize, aggregate, and compare heterogeneous quantitative measures.

With financial information as a basis, knowledge can be produced and diffused within the organization. Formal management control systems can be used to produce knowledge that makes it possible to exercise what is called "remote control". This has many advantages, but it can also be the cause of extensive problems (Johnson, 1992). This is particularly true when information is handled without an understanding of the actual situation. Or, as Marshall *et al.* (1997, p. 232) express it: "Information without knowledge of the context of the information is very dangerous."

In many business firms today, the traditional instruments for management control are perceived as giving a poor, and sometimes even wrong, picture of the business. Those with financial functions who are working with the information are simultaneously criticized for a lack of knowledge about the enterprise (Johnson and Kaplan, 1987). The explanation for this situation is many of the models and methods used in management control systems were generated for far less complex business firms than those we see today (Emmanuel, Otley and Merchant, 1990). Whereas earlier business firms worked with standardized products in a few markets, modern firms may be characterized by products with a large service component, adapted to specific customers, and available on many markets.

The information produced by traditional management control systems has been criticized for being too narrow and taking only the financial dimension into account. As a consequence of the extensive criticism, new methods have been developed. There is an emerging interest in alternative ways of measuring performance (Eccles, 1991; Kaplan and Norton, 1996), using operational, non-financial measurements and goals (Euske, Lebas and McNair, 1993).

Modern management control systems aim to strengthen the possibilities for organizational learning, as argued in Kaplan's and Norton's influential book, *Balanced Scorecard* (1996), which includes this among the many reasons for using a new management control system. Where traditional management control systems are criticized for being too oriented towards stability, modern management control systems contribute to a more dynamic situation. In other words, traditional management control systems generate and distribute information that helps us improve what we already know, whereas modern management control systems aim to make us learn new things (Simons, 1995). The new information generated by the measurements and methods of modern management control systems not only improves a business firm's efficiency, but also makes it possible to question established ways of doing things by creating a drive for effectiveness and improved learning.

Organizational learning through management control systems

Argyris and Schön (1978) described organizational learning as the process whereby members of an organization detect and correct errors that have originated from internal and external changes. For learning to be possible, existing beliefs and values must be questioned. Learning is the process through which learners discard knowledge and become able to acquire new responses and new mental (cognitive) maps. Knowledge acquisition can be said to be a product of both intentional searching and unintentional noticing (Huber, 1991). Argyris and Schön (1978) have defined two different types of organizational learning: single-loop and double-loop learning. Single-loop learning is the more instrumental way of learning, which changes an individual's actions without having an effect on the individual's more fundamental beliefs and values. Double-loop learning necessarily involves detection and correction of errors, but also assumes a change in fundamental beliefs and values. Double-loop learning "results in a change in the values of theory-in-use, as well as in its strategies and assumptions" (Argyris and Schön, 1996, p. 21).

Learning is stimulated both by environmental changes and internal factors in a complex and iterative manner. Learning presumes that organizational experiences are maintained and accumulated (Levitt and March, 1988). Argyris and Schön (1978, p. 19) asserted that "learning agents' discoveries, inventions, and evaluations must be embedded in organizational memory." According to Starbuck (1983) unlearning tends to be more difficult when rigid hierarchies are formed that insulate top management from reality, or when an organization relies on action generators such as a calendar or a budget. It is also more difficult in cases where organizations punish dissent and deviancy, buffer themselves from the environment, or use over-simplified perceptual categories to facilitate documentation and communication.

Information systems can support the processes of knowledge acquisition, information distribution, and information interpretation needed for organizational learning. Information systems can also be seen as a part of organizational memory, storing knowledge for the future (Huber, 1991; Walsh and Ungson, 1991). Information interpretation is the process for understanding (interpreting) information. Because individuals and groups have prior belief structures, any information received is automatically treated and shaped by these. The interaction between stored belief structures and interpretation has been shown to be critical to understanding how organizations learn. Greater learning occurs when more varied interpretation is developed. The more individuals learn in an organization, the more similar individual belief patterns become throughout the organization. But as individuals become more similar, organizational learning decreases, since there are fewer new and dissenting ideas from organizational participants (March, 1991).

It has been argued that management control systems, including accounting systems, can be discussed in terms of structures and processes (Anthony and Govindarajan, 1995). They are designed to achieve organizational control and are used by managers to assist them in decision-making, communication, motivating, coordinating, and integrating activities (Kloot, 1997). A firm's planning through the budget process is the basis (standard) on which information can later be created. This is done when the real outcomes are compared with the planned outcomes. Any discrepancies constitute information to which actors can respond by taking action. Cybernetic control systems like this are the foundation on which many management control systems are built (Beniger, 1986; Otley, 1994). Textbooks in the field often emphasize the positive effects of budgets, such as that they improve communication and the learning possibilities within a firm (Arwidi and Samuelson, 1991, p. 21).[1] An important distinction is often made between "tight" and "loose" budgetary control. This distinction affects the possibilities for learning. A tight budget is subject to strong central control, and specifies many details in regard to the budget object (Anthony and Govindarajan, 1995). Experience has shown that such control easily leads to withholding of information and a lack of communication, and risks manipulation of the distributed information (Merchant, 1981). The attitude towards the budget is much freer when there is loose control. Within an ideology of tight control, the budget is seen as a "commitment", something that has to be achieved. In a loose-control ideology, the budget is perceived as a "best estimate" of the future. The characteristics of loose control, such as an open attitude and goal orientation (instead of tight, rigid, and detail-oriented control), improve communication and organizational learning (Hopwood, 1973).

Traditional management control systems are commonly described in terms of formalized, financial control of the entire organization. The formalized structures and processes contribute to the stabilizing function of management control systems (Simons, 1995). However, the single-minded focus on financial matters is, as we have seen, being questioned by many. One of the points made is that the traditional information does not elicit what is most interesting in a company (McKinnon and Bruns, 1992). This debate has led to increased interest in the use of non-financial measurements and methods, such as qualitative and operational measurements,

---

[1]Recently the idea of a budget has come under strong criticism from many viewpoints. It is often emphasized that a budget is a very restricted tool for comparison. The turbulent economic environment makes it impossible to specify business well in advance (Wallander, 1994).

to improve organizational learning (Kaplan and Norton, 1996). There is growing awareness that an organization's management control systems can no longer view the organization as an isolated unit, but must take account of important actors in the environment (Shank and Govindarajan, 1994). More attention is being paid to customers and suppliers. Just as the traditional, one-sided, financially oriented budget provided good support for single-loop learning, so the multidimensional use of information in more modern management control systems improves the conditions for double-loop learning. Interaction with competent and demanding customers or good suppliers may help a firm grow. Better information about these important groups of actors improves the conditions for double-loop learning. *Ad hoc* reports also contribute to the conditions for double-loop learning, for information about "exceptions" supplements formal, periodic reports and makes knowledge more holistic (Hedberg and Jönsson, 1978).

Information becomes knowledge through a dialogue between actors, and that is why patterns – frequencies and forms of communication – are important (Luckett and Eggleton, 1991). The frequency of contact greatly affects the opportunities for organizational learning (the more, the better). But the channels of communication are also important. Different kinds of information need different kinds of channels (Daft and Lengel, 1984). Just as different organizations need different information, so network organizations need richer communication (Eccles and Nohria, 1992). Simple, standardized information can easily be transferred electronically, while more complicated information has to be handled through richer media, such as personal and verbal contacts. Personal contacts are of greatest importance when it comes to the transmission of tacit knowledge (Nonaka and Takeuchi, 1994). Relating this to the concepts of single and double-loop learning, it can be shown that the richer and more individual the contacts, the more they will improve the conditions for organizational double-loop learning.

Conclusion

At the same time as many global companies are changing from a hierarchical to a network-oriented structure, traditional management control systems are being criticized and new, modern methods of management control are evolving (Bromwich and Bhimani, 1994). Both of these trends focus on the desire to improve the conditions for organizational learning. Because different organizations need different information, the question arises whether network organizations with their greater need for coordination use a management control system that creates better opportunities for increased organizational learning. As we have seen, management control systems can be considered important tools for organizational learning. Modern management control systems create better conditions for organizational learning as they include more loosely controlled budgets, more non-financial and operational information, and more open and flexible dialogue. But do subsidiaries in a network organization use modern management control systems? Is their budget control looser than that in a traditionally organized company? Do they use non-financial and operational information for planning and follow-up? Do they use *ad hoc* reports, and do they differ from traditionally organized firms by having more frequent communication (dialogue) characterized by more content-rich, personal meetings?

Method

In order to answer the above questions and study the management control systems used by global companies, a questionnaire was constructed and mailed to the 200 largest (in terms of the number of employees) global companies in Sweden. A global company was defined as a company with its parent company in another country. The companies were identified through the register of Statistiska Centralbyråns Företagsregister (Statistics Sweden) in which, by law, every organization in Sweden is registered. The questionnaire was sent to the financial officer of each company, and 140 of the 200 questionnaires were returned. This gave a response rate of 70%. Even though this could be considered satisfactory, it is also important to look at the companies that did not respond. Of the 60 companies that did not respond, seven informed us on their own initiative that they were not able to return the questionnaire because of a lack of time, change programs or mergers. The non-response analysis considered the companies' size, business sector, and the national location of the parent companies, and established that the remaining 53 companies who had not answered the questionnaire did not differ in any significant respect from the 140 who had responded. No differences could be found between those companies that returned the questionnaire early and those that returned it later, after a reminder.

The global companies studied can be divided into two groups: 1) those within organizations consisting of a group of companies, some of which had particular responsibilities (e.g. "centers of excellence"); and 2) those within more traditional organizations where such units do not exist. Bartlett and Ghoshal (1989) would categorize the companies in the first group (structured as an integrated network) as transnational organizations, but we have chosen to call them *network organizations*. The second group, operating within less integrated and more *traditionally* structured organizations, we have called *traditional organizations*. Of the 140 companies studied, 95 are network organizations and 45 are traditional organizations.

The parent companies were located in Europe (99 companies), the U.S.A. (36 companies), Japan (4 companies), and Kuwait (only 1 company). The European companies were located in Germany (18 companies), Finland (13 companies), Denmark (13 companies), Great Britain (12 companies), France (9 companies), The Netherlands (9 companies), Switzerland (8 companies), Norway (8 companies), Belgium (5 companies), and a company each in Luxembourg, Italy, Austria, and Liechtenstein. The majority of the companies (66%) have more than 200 employees, with the average number of employees being 450. The companies represent different business sectors.

The questionnaire focused on perceptions of management control systems and coordination activities in the global company. One part dealt with the aspects discussed in this article, while other parts related to other aspects of management control systems. Earlier research has shown the importance of perceptions of management control system (Pettigrew, 1973; Bariff and Galbraith, 1978; Swieringa and Weick, 1987, p. 296; Markus and Pfeffer, 1983; Merchant, 1985). Perceptions of the control system influence the actions of members of the organization. People pay attention to what is measured, and it becomes part of the perceived, inter-subjective reality of the organization (Ridgway, 1956; Eccles, 1991).

Care was taken when constructing the questionnaire that all questions were clear and straightforward in four important respects: they used simple language, referred to common concepts, set manageable tasks, and drew on widespread information (Converse and Presser, 1986).

Empirical data

The use of the budget

We anticipated that companies within network organizations would have relatively loose budget control since learning and knowledge transfer are supposed to be more important for them. The idea is, as discussed earlier, that the less detailed the budget, the greater the possibilities for learning. Of 140 responses, 134 were companies using a budget system. This means that only six companies (four within network organizations, two within more traditional organizations) were not using a budget at all. But what was the situation for those that did use budgets?

Table 13.1 indicates that, contrary to what would have been expected, companies within network organizations are characterized by tighter control than traditionally organized companies. The majority of the companies within network organizations, 75 of 91 (82.4%), considered their budget to be characterized by many details, compared with roughly 72% in traditionally organized companies.

TABLE 13.1
Levels of details and amount of budget instructions from central unit

	Network Organization	Traditional Organization	Total
Many details	75 (82.4%)	31 (72.1%)	**106 (79.1%)**
Many instructions	72 (79.1%)	31 (72.1%)	**103 (76.9%)**
Few instructions	19 (20.9%)	12 (27.9%)	**31 (23.1%)**
Total	**91 (100%)**	**43 (100%)**	**134 (100%)**

In the empirical study, the character of the budget was also operationalized in the form of a question about the number of clear instructions issued by the superior unit in the organization. A situation with few clear instructions attests to decentralized, looser control, as exemplified in the budget philosophy quoted by Anthony and Govindarajan (1995, p. 499): "I hire good people, and I leave them alone to do their jobs." Responses to the question regarding instructions from the superior unit are summarized in Table 13.1:

Table 13.1 clearly shows that companies in a network organization are working in a budget situation characterized by many instructions from the central unit. Almost 80% of the companies in a network organization are characterized as working with centralized instructions to a great extent. The corresponding number for traditionally organized companies is 72%.

The statistical data illustrate that, in regard to both the level of detail and instructions, the situation is other than expected. Loose budget control, which creates a climate favorable to organizational learning, is not dominant among the network companies. Instead they seem to be characterized by tight budget control.

The use of non-financial measurements

The conditions for organizational learning are also enhanced by the use of information that is not strictly financial. To what extent is such information used in reports from the subsidiary to the central unit? Do companies in a network organization, where learning and the transfer of knowledge is expected to have higher priority, use non-financial measurements more often?

For the sake of simplicity, we will focus on those groups of companies that do not use non-financial measurement at all and those that use non-financial measurements to a great extent. For the present, we will ignore the large group which answered "to some degree".

Companies within a network organization make slightly more use of non-financial measurements (see Table 13.2). Nearly 17% of the

TABLE 13.2

The use of non-financial measurements

	Network Organization	**Traditional Organization**
Not at all	17 (17.9 %)	11 (24.4%)
To some degree	62 (65.3%)	30 (68.2%)
To a great extent	16 (16.8%)	4 (9.1%)
Total	**95 (100%)**	**45 (100%)**

companies in such organizations responded "to a great extent," as compared with the more traditionally organized companies, where only 9% gave this response. At the other extreme, the companies within traditional organizations more often responded with "not at all", when asked if they used non-financial measurements. The percentage of companies within a traditional organization not using non-financial measurement was 24%, as compared with 18% of the network companies. The differences between the two types of organizations are smaller than we expected.

The use of operational measurements

New measurements can be defined more precisely in terms of the extent to which they are operational. As mentioned earlier, management control systems are often criticized for being too much oriented towards the information needs of external interests (Walther *et al.*, 1997). This has resulted in information that is too aggregated, and that risks overlooking aspects that are of interest to internal actors. This is a problem if management control systems are considered a tool for learning and development. Operational measurement is needed to improve and supplement traditional management control systems. Such measurements can take the form of standards for processes, suppliers, and customers. In a management control system these standards can also serve as positive disturbing factors to stimulate organizational learning (Hedberg and Jönson, 1978). To what extent are operational measurements used in the studied companies (see Table 13.3)?

In this case, too, the extreme categories "Not at all" and "To a great extent" are the most interesting. The responses indicate that operational measurements are more frequent in companies within network organizations. Almost 14% of these companies report such measurements to their superior unit. Corresponding numbers for companies in traditional organizations are just above 9%. The category of companies not using operational measurements at all is smaller for the network companies,

Carin B. Eriksson and Jan Lindvall

TABLE 13.3
The use of operational measurements

	Network Organization	Traditional Organization
Not at all	27 (28.7%)	15 (33.3%)
To some extent	54 (57.5%)	26 (59.1%)
To a great extent	13 (13.8%)	4 (9.1%)
Total	**94 (95) (100%)**	**45 (45) (100%)**

29% as compared with the 32% of more traditionally organized companies.

The use of ad hoc *reports and personal meetings*

As mentioned earlier, an interruption in the formalized cycle of control can be of great importance in improving organizational learning. In such cases, *ad hoc* reports are essential. *Ad hoc* reports, which fall outside the normal cycle of reports, improve the possibility of learning more about things that are not included in standard reports.

To what extent did the studied companies use *ad hoc* reports? How many times has the superior unit requested *ad hoc* reports over the last 12 months? In view of the fact that network companies have greater need of well-developed conditions for organizational learning, they could also be expected to use *ad hoc* reports more often. The empirical data (see Table 13.4) indicate that the group of companies within network organizations supplied *ad hoc* reports an average of 20 times over the past 12 months. The average number of *ad hoc* reports supplied by more traditionally organized companies was nine.

An important objective of a management control system is to provide a base for communication and dialogue within the company. How many times a year does the company in Sweden meet their superior unit to discuss different topics? Table 13.4 shows the average number of such meetings in the two types of organization.

TABLE 13.4
Average number of ad hoc *reports and meetings*

	Network Organization N=95	Traditional Organization n=45
Average number of *ad hoc* reports over past year	20.0	9
Average yearly number of meetings	7.0	5.4

236

On average, the 95 companies within network organizations meet their superior units more often than the companies in traditional organizations. The average for the network companies is seven times a year, while for the traditionally organized companies it is five times a year.

When thinking about communication, it is important to consider not only the frequency of the contact but also the richness of the information channels (Daft and Lengel, 1986). Personal contact is the richest method of communication, followed by telephone calls, electronic mail, memos, letters, and bulletins. The channels vary in content and in how much of the communication is a matter of routine. The most content and the least routine are possible in personal meetings. What is the situation in the companies studied? Table 13.5 shows the amount of contact by telephone and electronic mail.

TABLE 13.5
Channels of information

	Telephone		Electronic mail	
	Network	**Traditional**	**Network**	**Traditional**
Often	79 (84.9%)	38 (84.4%)	71 (77.2%)	20 (44.4%)
Rarely	14 (15.1%)	7 (15.6%)	21 (22.8%)	25 (55.6%)
Total	93 (100%)	45 (100%)	92 (100%)	45 (100%)

The statistical data indicates that personal contacts between superior and subordinate units are more frequent in network companies (see Table 13.4) and that these companies make greater use of electronic mail as a channel for information (see Table 13.5). While 77% of the network companies are often in contact with their superior unit by electronic mail, the corresponding number for traditionally organized companies is only 44%. The more frequent use of this relatively poor channel of information has to be seen in combination with the more frequent use of personal contacts in the network organizations. The personal communications are supplemented with poorer forms of information. The correlation between these two forms of information indicates a covariance coefficient of 0.23. No important differences could be observed in regard to contacts by telephone.

Concluding discussion

The point of departure for this article was that new organizational structures emerging in highly internationalized firms require greater cooperation, communication and information. This development does not merely impose new demands, it also provides better possibilities for organizational learning. At the center of the new demands for information are management control systems. The new structure – the network structure – necessitates a greater emphasis on control of information from subsidiaries. Many of the companies studied have the form of network organizations. Almost 68% (95 of 140) are organized in such a way that units are given special responsibilities. These units may be called centers of excellence, international product centers or shared service centers. The traditional tools for management control systems are often perceived as rigid, allowing little scope for organizational learning. Since modern management control systems aim to strengthen the possibilities for organizational learning, one would expect network organizations to use more modern management control systems.

Textbooks about management control systems often emphasize the positive effects of budgets, such as improving communication and learning possibilities within the firm (Arwidi and Samuelson, 1991, p. 21). Recently, however, the concept of a budget has been strongly criticized on the grounds that the turbulent economic environment makes it impossible to predict business well in advance, and thus restricts the budget's usefulness as a tool for comparison (Wallander, 1994; CFO Europe, 1998). Despite this criticism, the overwhelming majority of the companies studied do use budgets and only six of the 140 companies do not. A firm's planning through the budget process provides the basis (standard) for the generation and interpretation of later information, and subsequent action. When it comes to organizational learning, there is an important distinction between "tight" and "loose" budget control. Contrary to expectation, the study showed that the network organizations are using budgets less adapted to organizational learning than those of traditional organizations. The budgets in the network organizations are often concerned with various kinds of instructions and details (see Table 13.1). According to previous research (Merchant, 1981) such "tight" control, with many levels of detail in the budget, may lead to withholding of information and risks manipulation of the distributed information. In a loose-control ideology, the budget is perceived as a "best estimate" of the future. The characteristics of loose control, such as an open attitude and goal orientation (compared with tight, rigid, detail-oriented control) should improve communication and organizational learning (Hopwood, 1973). But neither network organizations nor traditional organizations are using loose budget control.

Another characteristic of the new, more developed management control is that information is perceived as broader than merely the traditional financial dimension. New dimensions, such as non-financial and operational information, are becoming more important. So are *ad hoc* reports, originating outside the normal, formalized, regular cycle of reports. Contacts between actors (referring to the informal part of management control systems) are also supposed to be more frequent and richer in information in network organizations as their need for good systems for organizational learning is greater. The results of the empirical study does not prove that network organizations are using non-financial and operational measurements more often. They are using *ad hoc* reports more often, and they have more frequent personal contacts (see Tables 13.2 and 4), *but* the differences between the two types of organizations are small. In fact, the indications that they are using a traditional management control system are stronger than the indications for a modern management control system.

In conclusion: there is no strong evidence that the network organizations studied are using more modern management control systems. Indeed, we can see that they are using a modern management control system in some areas but not in others. It is possible that both the trend toward organizing in networks and the changes in modern management control systems are too recent to have made much of a difference yet. Changing ways of thinking and acting within an organization takes time, and the new ideas may not yet have broken through.

Other important questions remaining to be discussed in a later article include the effect of a company's size, traditions, and the "nationality" of its parent company. National, social, and cultural differences may, of course. influence how control of subsidiaries is valued, as Bartlett and Ghoshal (1989) show in their discussion of the differences between the triad's regions.[2] On a broader level, diversity may also be discussed with reference to disparities between different national systems, such as the shareholder systems found in the U.S.A. and Great Britain and stakeholder systems such as those found in Germany and Japan (Stewart *et al.*, 1994).

References

ANTHONY, R. J., and GOVINDARAJAN, V., 1995, *Management Control Systems*, 8th edition, Chicago: Irwin.

ARGYRIS, C., and SCHÖN, D. A., 1978, *Organizational Learning*, Reading, MA: Addison-Wesley.

ARGYRIS, C., and SCHÖN. D. A., 1996, *Organizational Learning II: Theory, Method and Practice*, Reading, MA: Addison-Wesley.

ARWIDI, O., and SAMUELSON, L. A., 1991, *Budgetering i industriföretagets styrsystem*, Stockholm: Mekanförbundets Förlag.

BARIFF, M. C., and GALBRAITH, J. R., 1978, Intraorganizational Power Considerations, for Designing Information Systems, *Accounting, Organizations and Society*, 3, 1, pp. 15–27.

---

[2]Ohmae (1985) discusses the importance of having activities in the three regions: Europe, North America, and Japan.

Carin B. Eriksson and Jan Lindvall

BARTLETT, C. A., and GHOSHAL, S., 1989, *Managing Across Borders. The Transnational Solution*, Boston, Mass.: Harvard Business School Press.

BENIGER, J. R., 1986, *The Control Revolution: Technological and Economic Origins of the Information Society*, Cambridge, Mass.: Harvard University Press.

BROMWICH, M., and BHIMANI, A., 1994, *Management Accounting Pathways to Progress*, London: CIMA.

BUCKLEY, P. J., and CASSON, M. C., 1998, Models of Multinational Enterprise, *Journal of International Business Studies*, 29, 1, pp. 21–44.

CFO Europe, 1998, All for one and one for all, vol. 1, no. 3, July/August, pp. 12–18.

CONVERSE, J.M., and PRESSER, S., 1986, Survey Questions: Handcrafting the Standardized Questionaire, *Quantitative Applications in the Social Sciences*, p. 63, London: Sage.

DAFT, R. L., and LENGEL, R., 1986, Organizational information requirements, media richness and structural design *Management Science*, 32, pp. 554–571.

DUNNING, J. H., 1998, Location and the Multinational Enterprise: A Neglected Factor? *Journal of International Business Studies*, 29, 1, pp. 45–66.

ECCLES, R. G., 1991, The Performance Measurements Performance Manifesto *Harvard Business Review*, Nov-Dec., pp. 149–161.

ECCLES, R. G., and NOHRIA, N., 1992, *Beyond the Hype. Rediscovering the Essence of Management*, Boston, Mass.: Harvard Business School Press.

EMMANUEL, C., OTLEY, D., and MERCHANT, K., 1990, *Accounting for Management Control*, 2nd edition, London: Chapman and Hall.

EUSKE, K. J., LEBAS, J., and McNAIR, C. J., 1993, Performance Management in an International Setting *Management Accounting Research*, 4, pp. 275–299.

FORSGREN, M., 1988, *Managing the Internalization Process*, London: Routledge.

FRISO DEN HERTOG, J., 1978, The Role of Information and Control Systems in the Process of Organizational Renewal: Roadblock or Road Bridge? *Accounting, Organizations and Society*, vol. 9, 2, pp. 125–135.

HEDBERG, B., and JÖNSSON, S., 1978, Designing Semi-Confusing Information Systems for Organizations in Changing Environments *Accounting, Organizations and Society*, vol. 3, 1, pp. 47–64.

HEDLUND, G., 1993, Assumptions of Hierarchy and Heterarchy with Applications to the Management of the Multinational Corporation, in Ghoshal, S., and Westney, D.E., *Organization Theory and the Multinational Corporation*, New York: St Martin's Press.

HOPWOOD, A., 1973, *An Accounting System and Managerial Behaviour*, London: Saxon House.

HUBER, G. P., 1991, Organizational Learning: The Contributing Processes and the Literatures *Organization Science*, 2, pp. 88–115.

JOHNSON, T. H., and KAPLAN, R.S., 1987, *Relevance Lost. The Rise and Fall of Management Accounting*, Boston, Mass.: Harvard Business School Press.

JOHNSON, T. H., 1992, *Relevance Regained. From Top-Down Control To Bottom-Up Empowerment*, New York: The Free Press.

KAPLAN R. S., and NORTON, D. P., 1996, *Balanced Scorecard, Translation Strategy into Action*, Boston, Mass.: Harvard Business School Press.

KLOOT, L., 1997, Organizational Learning and Management Control Systems: Responding to environmental change *Management Accounting Research*, 8, pp. 47–73.

LEVITT B., and MARCH, J. G., 1988, Organizational Learning, *Annual Review of Sociology*, 14, pp. 319–340.

LUCKETT, P. F., and EGGLETON, I. R. C., 1991, Feedback and Management Accounting: A Review of Research into Behavioural Consequences *Accounting, Organizations and Society*, 16, 4, pp. 371–394.

240

MARCH, J. G., 1991, Exploration and Exploitation in Organizational Learning, *Organization Science*, 2, pp. 71–87.

MARKUS, M .L., and PFEFFER, J., 1983, Power and the Design and Implementation of Accounting and Control Systems *Accounting, Organizations and Society*, 8, pp. 67–85.

MARSHALL, C., PRUSAK, L., and SHPILLBERG, D., 1997, Financial Risk and the Need for Superior Knowledge Management, in Prusak, L., *Knowledge in Organizations*, Boston, Mass.: Butterworth-Heinemann.

McKINNON, S. H., and BRUNS, W. J., 1992, *The Information Mosaic. How Managers Get the Information They Really Need*, Boston, Mass.: Harvard Business School Press.

MERCHANT, K. A., 1981, The Design of the Corporate Budgeting System. Influences on Managerial Behaviour and Performance *The Accounting Review*, LVI, 4, pp. 813–829.

MERCHANT, K. A., 1985, Organizational Controls and Discretionary Program Decision Making: A Field Study *Accounting, Organizations and Society*, 10, pp. 67–85.

NONAKA, I., and TAKEUCHI, H., 1995, *The Knowledge-Creating Company. How Japanese Companies Create the Dynamics of Innovation*, New York: Oxford University Press.

OHMAE, K., (ed.), 1995, *The Evolving Global Economy. Making Sense of the New World Order*, Boston, Mass.: Harvard Business Review.

OTLEY, D. T., 1994, Management Control in Contemporary Organizations: Towards a Wider Framework, *Management Accounting Research*, 5, pp. 289–299.

PASTERNACK, B. A., and VISCIO, A .J., *et al.*, 1998, *The Centerless Corporation. A New Model For Transforming Your Organization for Growth and Prosperity*, New York: Simon and Shuster.

PETTIGREW, A. M., 1973, *The Politics of Organizational Decision Making*, London: Tavistock.

RIDGWAY, V. F., 1956, Dysfunctional Consequences of Performance Measurements, *Administrative Science Quarterly*, 1, 2, pp. 240–247.

SHANK, J. K., and GOVINDARAJAN, V., 1993, *Strategic Cost Management. The New Tool for Competitive Advantage*, New York: The Free Press.

SIMONS, R., 1990, The Role of Management Control Systems in Creating Competitive Advantage: New Perspectives, *Accounting, Organizations and Society*, 15, pp. 127–143.

SIMONS, R., 1995, *Levers of Control. How Managers Use Innovative Control Systems to Drive Strategic Renewal*, Boston, Mass.: Harvard Business School Press.

SJÖSTRAND, S. E., 1997, *The Two Faces of Management, The Janus Factor*, London: Thomson Business Press.

STARBUCK, W., 1983, Organizations as Action Generators, *American Sociological Review*, 48, No. 1, pp. 91–101.

STEWART, R., BARSOUX, J-L., KIESER, A., GANTER, H-D., and WALGENBACH ????, (1994) *Managing in Britain and Germany*, New York: St. Martin's Press.

SWIERINGA, R. J., and WEICK, K. E., 1987, Management Accounting and Action, *Accounting, Organizations and Society*, vol. 12, 3, pp. 293–308.

WALLANDER, J., 1994, *Budgeten ett onödigt ont*, Stockholm: SNS Förlag.

WALSH, J. P., and UNGSON, G. R., 1991, Organizational Memory, *Academy of Management Review*, 16, No. 1, pp. 57–91.

WALTHER, T., JOHANSSON, H., DUNLEAVY, J., and HJELM, E., 1997, *Reinventing the CFO. Moving from Financial Management to Strategic Management*, New York: McGraw-Hill.

Author Index

Subject index